I0824134

ALSO BY KORY STAMPER

Word by Word: The Secret Life of Dictionaries

TRUE COLOR

TRUE COLOR

The Strange and Spectacular Quest
to Define Color—from Azure to Zinc Pink

Kory Stamper

ALFRED A. KNOPF
New York
2026

A BORZOI BOOK
FIRST HARDCOVER EDITION PUBLISHED
BY ALFRED A. KNOPF 2026

Published by Alfred A. Knopf, a division of Penguin Random House LLC, 1745 Broadway, New York, NY 10019.

"What Exactly Is 'Cerise'? The Complicated History of Color Definitions" was originally published in *Slate,* 2014.

Excerpts from the Godlove correspondence published with permission from the Godlove family.

Library of Congress Cataloging-in-Publication Data
Names: Stamper, Kory, author
http://id.loc.gov/authorities/names/n2016042151
http://id.loc.gov/rwo/agents/n2016042151
Title: True color : the strange and spectacular quest to define color— from azure to zinc pink / Kory Stamper.
Description: First hardcover edition. | New York, NY : Alfred A. Knopf, 2026. | "A Borzoi book." | Includes bibliographical references and index.
Identifiers: LCCN 2025011260 (print) | LCCN 2025011261 (ebook) | ISBN 9781524733032 hardcover | ISBN 9781524733049 ebook
Subjects: LCSH: Colors, Words for http://id.loc.gov/authorities/subjects/sh85028703 | Lexicology http://id.loc.gov/authorities/subjects/sh85076359
Classification: LCC P305.19.C64 S73 2026 (print) | LCC P305.19.C64 (ebook) | DDC 535.601/4—dc23/eng/20250625
LC record available at https://lccn.loc.gov/2025011260
LC ebook record available at https://lccn.loc.gov/2025011261

penguinrandomhouse.com | aaknopf.com

Printed in the United States of America
3rd Printing

The authorized representative in the EU for product safety and compliance is Penguin Random House Ireland, Morrison Chambers, 32 Nassau Street, Dublin D02 YH68, Ireland, https://eu-contact.penguin.ie.

For Margaret

The living room: I want it to be a *soft green*. Not as blue-green as a robin's egg, but not as yellow-green as daffodil buds. Now, the only sample I could get is a little too yellow, but don't let whoever does it go to the other extreme and get it too blue. It should just be a sort of *grayish-yellow-green*.

—Myrna Loy as Muriel Blandings,
Mr. Blandings Builds His Dream House (1948)

scotch gray *n, often cap S* : a grayish yellow green that is yellower and paler than average sage green or palmetto and greener and duller than mermaid

—*Webster's Third New International Dictionary of the English Language, Unabridged* (1961)

Contents

TRUE
COLOR

Preface

When I was hired by Merriam-Webster in 1998, it was ostensibly to revise the Big Book, *Webster's Third New International Dictionary, Unabridged*. The *Third,* or W3 as it's called in the office, was released in 1961 and it made a splash. A dictionary written for the nuclear age, it's 2,662 pages of six-point type, ten pounds of knowledge stuffed between two buckram-covered boards, the result of tens of thousands of editorial hours by more than a hundred in-house editors and two hundred outside experts. It's a nonpareil of twentieth-century American lexicography, notable for its almost scientific, systematic approach to what belongs in a dictionary and how the words inside it should be defined. Every modern American dictionary that you've consulted owes something to the *Third*—even if that something is that the dictionary you're consulting is *not* the *Third*.

I say "ostensibly" because every editor who had been hired at Merriam-Webster since the mid-1970s had been hired to revise the *Third,* but no work had been undertaken on a full revision of the beast since its publication. New words had been added by means of an Addenda Section slapped into the front of the book—editing this was, in fact, my first defining job as a lexicographer—but the original *A–Z* remained relatively untouched. Nonetheless, every editor who crept quietly up the stairs to the editorial floor and took their creaky seat was trained using all the same techniques and style sheets used to write the *Third*. When you wrote practice definitions, they were graded on the curve of the *Third*. Not sure whether the usage label for this archaic term for the female genitals should be *"dialectal, chiefly England"* or *"chiefly British dialect"*? Lift up thine eyes to the *Third*. Every other dictionary we wrote, from the mass-market paperback to the *Collegiate*, owed its existence to the *Third*. This was done not because we love tradition, or because the *Third* was such a big deal that we could never deviate from it, but in anticipation of the big day when the stars would align and the market would be ready for another big ol' buckram-covered computer-monitor stand. When that day came, every lexicographer on staff would have to drop everything and swan dive gracefully into the complexities of the Big Book.

In 2010, almost fifty years after its initial publication, the stars aligned. We bounced on our toes, took a deep breath, and began the unending work of revising the *Third* into a new unabridged dictionary.

That is when my love affair with color began.

One of the jobs of a lexicographer is to proofread dictionary entries. This task is more than simply reading each word in an entry: Proofreading dictionary entries includes finger stepping your way through each etymology to make sure that the dot underneath that *h* is supposed to be there and isn't just fly poop, verifying that the angle brackets which introduced an example sentence were in roman and not accidentally italicized (because even though no one but a professional typesetter would ever notice it, it's still wrong), and checking that the initial colon which starts that six-point-type definition is in boldface and surrounded by exactly one space on either side—and not an en or em space, either. It is brain-melting tedium on a microscopic scale, and I enjoy it immensely.

I had been asked to proofread some changes and new additions we had made to entries in the letter *B,* and make sure that they were being translated properly for the website that housed the new-but-in-progress *Unabridged Dictionary*. I started at "Beaufort scale" and began plodding my way through the word list, noting whenever an etymological character was rendered incorrectly, a link to a table was broken, or a fat-fingered extra space in the code resulted in a blank page. This is exactly what I had trained for: to be able to spot minute errors in transliterating the *Third* without needing to study the page.

The *Third* has a very distinctive defining style. Dry, impersonal, a little robotic to the point of hilarity. Gone are whimsical, snappy definitions reminiscent of Samuel Johnson, the eighteenth-century lexicographer who famously defined "oats" as "a grain, which in England is generally given to horses, but in Scotland supports the people." The *Third* was no-nonsense and all business, full of "the state or condition ofs" and "of, relating to, or characterized bys." No narrative interest, no silly puns, no personality whatsoever—just exhaustive definitions that look as if they were written by very articulate alien visitors to earth. That is the ultimate goal of every definition in the *Third*. This is a dictionary that expanded the food-related meaning of "hunger" into three separate definitions, one of which was "an uneasy sensation occasioned normally by the lack of food and resulting directly from stimulation of the sensory nerves of the stomach by the contraction and churning movement of the empty stomach." This is a dictionary that defined the word "fish stick" as "a stick of fish."*

* Originally, at any rate. It is now defined a little more sensibly in the online *Unabridged Dictionary* as "a small, elongated, breaded fillet of fish." *Requiescat in pace,* bad definition.

So you will perhaps understand why the entry for "begonia" caught my eye. Two botanical definitions, as expected, with all hyperlinks working and all acute accents in the etymology intact. Then I came to the third definition in the entry:

> **3 -s :** a deep pink that is bluer, lighter, and stronger than average coral (see CORAL 3b), bluer than fiesta, and bluer and stronger than sweet william — called also *gaiety*

I blinked. I read it again. The definition sure fit the form of a definition for the *Third,* but it was utter and complete nonsense. Here was a color—"a deep pink"—compared color-wise to a handful of things that I was pretty sure were not colors. "Sweet william" I recognized as the name of a flower, so maybe there was also a color called sweet william that was based on the color of the flower. I googled "sweet william" and the first picture that showed up was of a hedge of flowers that were white, bright magenta, dull purple, middle-of-the-road pink, and even a dusty, blue-tinged lilac. Which of those sweet williams was "sweet william"? This was not an auspicious beginning. "Fiesta" I recognized as a party—mariachi bands and fireworks and piñatas. There is no color associated with a fiesta, apart from maybe *all* the colors, as befits a party. Maybe this "fiesta" was referring to the original color of Fiestaware, the brightly colored tableware from the 1940s that every grandma in the American West had a set of in the cabinet. But I was still stumped. Being neither from the 1940s nor a grandmother in the American West, I had to default to what I knew: "Fiesta" was a big ol' party. At least, I reasoned, I recognized "coral" as a color. But what the hell was "average" coral? "Average" implied the middle of a range, but what were the two ends of this particular range? Light and dark? Dull and bright? Orange and purple? Was "average" here referring more to the top of the bell curve—the most common color in a group of very similar colors that we call coral? Are there even enough corals to make a group of colors all called coral? I began wondering if there was "excellent coral," or "moderately disappointing coral."

There'd be an easy way to figure this all out. I whumped open my desk copy of the *Third* and began looking for the inevitable color chart that would explain the difference between "average coral" and "fiesta." There is a two-page color plate near the entry for "color" that includes exactly zero color chips named "average coral" or "fiesta." In fact, it contained no color chips at all: just a picture of the visible spectrum and a weird, multicolor blob from two sides.

Mine not to reason why, mine but to proofread and die: I clicked on the hyperlink at "see CORAL 3b" to make sure it worked, and was greeted by this:

> **3 :** something bright red in color: such as
>
> **a :** a bright-reddish ovary (as that of a lobster or scallop); *also* : the cooked roe of a lobster
>
> **b :** a variable color averaging a deep pink that is yellower and duller than fiesta or begonia and yellower and darker than sweet william
>
> **c of textiles :** a strong pink that is yellower and stronger than carnation rose, bluer, stronger, and slightly lighter than rose d'Althaea, and lighter, stronger, and slightly yellower than sea pink

Yes, there's the mirror of "begonia" in sense* 3b, but what fresh VistaVision glory is sense 3c! No longer words on a page, I was tum-

* Congratulations: You've encountered your first bit of dictionary jargon! "Sense" here refers to the numbered definitions inside a dictionary. Impress your boss/friends/pets with your newfound knowledge.

bling in and among those color names, bouncing around between sense memory and fantasy, utterly transported. *Carnation rose!* A close-up of the Carnation evaporated milk can popped to mind, that red-and-white label with the blue lozenge proclaiming that what was inside was "MILK," and scattered under the red banner, those eponymous flowers, in dark red, white, and what must be carnation rose. *Rose d'Althaea!* With that floral first name and that surname a jumble of extra vowels and strange capitalization, this is the color named after Scarlett O'Hara's well-to-do cousin from the French side of the family, the side not mentioned in *Gone with the Wind,* the cousin who attended cotillion in a silk dress the color of unopened magnolia blossoms and the next year abandoned Dixie and family to marry some damned Yankee.

But it was "sea pink" that undid me. The utter ridiculousness of it: the "sea," which I had understood to be a variety of blues, greens, grays, and other assorted not-red colors; and "pink," that Barbie doll, Pepto-Bismol, Material Girl, definitely not-blue color. *Sea pink.* I placed my hands very gently on my desk, then mouthed the words "sea pink" to myself and hoped that I could suppress the riptide of laughter which was sucking me—a team player who worked in a silent office—under. "Sea pink," I mouthed again, and this time the laughter hitched a ride with the *s* of "sea pink" and seeped out of me until I sounded like a leaky tire bouncing down a road of *p*'s and *k*'s.

The co-worker who sat in the cubicle behind me slammed his newspaper shut and stomped off to find a non-sibilant corner where he could continue to read without listening to me chuckle like a loon. I wanted to stand up and yell after him, by way of explanation, "I grew up fifteen hundred miles from an ocean! I didn't know the sea was *pink*!"

From that day forward, my workday breaks were filled with a

Technicolor romp through the *Third*. Tired of proofreading for a bit, I'd pull up the beta website for the unabridged and wiggle my fingers over the keyboard in invocation. Let's start at "aqua" this time, I'd say, and then I'd follow all the other colors listed in the definition through the dictionary and learn that "robin's-egg blue" is two different colors; that "cobalt" wasn't just blue, but could also be red, yellow, green, or violet; that "mallard" isn't a color but "duckling" is.

Part of what was so entrancing about these definitions was that they had *a voice*. The *Third* was carefully designed not to have any voice apart from the Corporate One. These color definitions—compared with the staid and rigid style of the rest of the *Third*—were as flashy as an entire team of cabaret dancers. They were unlike any other color definitions in any of our dictionaries. In fact, they were unlike any other color definitions, period. I had been involved in the creation of dozens of dictionaries at that point and had even written definitions for colors before. The idea was to aim in a spectral direction: "a bright red," "a moderate blue." Where in the bluer, yellower, slightly stronger hell had these extensive definitions that referenced "sea pink" and "copen" and thousands of other colors that I, a woman whose job was to read everything within reach and had never heard of before, come from?

The answers to these questions lay in piles of correspondence moldering in the Merriam-Webster basement; in family papers tucked safely away in archival boxes; in neatly typed notes in corporate archives; in long-out-of-print books; in government documents and reports to and from Congress; in fabric swatches and chemistry equations. Each one of these definitions is the crystalline, whittled-down result of dozens of overlapping concerns: the reach of science in the twentieth century, the commoditization of color, the scientism of American society and the inevitable backlash, the place of dictionaries in our cultural consciousness, the governmental foray into standardization, the insufficiency of language to describe the abstract, the drudgery of typesetting, the impact of war, the dangers of dye works, the power of love, and the ineffability of sea pink.

Like all good stories, it starts in the middle.

Chasing Rainbows

It's human nature to categorize. It's one of the easiest ways to make sense of the dizzying experience of being on this planet, and it's proven useful as we have made our way through history. This large furry mammal with the roar and the stripes and the long teeth will kill me, we say, while the smaller version of that furry mammal will merely live in my home and knock things off my shelves while it stares me straight in the face.

One handy marker we've used in categorizing things is color. It's such an elementary marker that it's also one of the first categories we teach our children. Panthers are black, pumas are brown, and tigers are? "Orange and black!" our preschoolers holler, and we beam with pride. (We will wait until they are in elementary school before we tell them that tigers can *also* be white and black, like zebras.) Color is so integral to our experience as humans that we can't conceive of a world without it. It's as much a part of our lives as air, water, and taxes. And it is maddeningly, beguilingly slippery.

Color is so maddening, so beguiling, because what we think we know about color contradicts how we experience it. Rainbows occur only when physical conditions are right—water particles suspended in the atmosphere at the right density, cloud cover broken enough to allow light to reach those water droplets, eyeballs turned skyward to perceive—but try to grab a rainbow, and nothing's there. Shine a red light at a wall and overlap it with a green light, and you get yellow light; slap some red paint on a wall and overlap it with green paint, and you get mud (assuming the red and green paints are both wet, and that you're mixing them on the wall well, and that the wall is nonabsorbent, and that the paint is not glossy, and and and). White and black, which we are accustomed to thinking of as *not colors,* are actually all the colors of the visible spectrum either entirely present and bouncing back at your retinas (white) or entirely absent and being wholly, greedily absorbed (black). We pass our cell phones around and look at a blurry picture of a striped dress that a stranger posted to the internet, and then holler at the one person at dinner who says that they think the dress is blue and black, because *oh my God, it's obviously gold and white, what is wrong with you, are you color-blind??* We are so immersed in color that when someone has even the slightest impairment in the way they see color, we say they are color-*blind*.

The science of color is also maddening for the same reason. Color, the physicist tells us, is a trick of light. Certain wavelengths of light from the visible spectrum* are reflected, refracted, or absorbed by the things they come into contact with: the ball on the floor, the floor, the wall behind the ball, the particles of dust in the air between you and the ball, the pollutants and water vapor in the atmosphere between the sun and you. Simple. Up pops the biologist to say that it's not *quite* that simple, because without the right combination of photoreceptors in the retina, a healthy retina, a solid optic nerve, a clear lens and healthy cornea, yadda yadda yadda—without the complex mechanisms of the human eye, we wouldn't perceive any of that light as color. And before the physi-

* *Actually* from the full spectrum of electromagnetic radiation, but most people are just interested in the visible spectrum, poor slobs.

cist can argue, a neurologist peers around the corner to say that color is a sensation generated by our brains to interpret the neural signals from the retina caused by light, and so, really, color can't happen independent of human cognition. You turn to leave and a psychologist blocks your exit and, as they push you back into the room and the horror-film music begins, tells you that the psychophysical sensation known as color is dependent upon the other color sensations your brain is interpreting at the same time, the language you assign to that sensation, and whether you are synesthetic. But how can you know, you say, voice rising into a shrill, that all of this works the same way in each person? They crowd in and you begin to hyperventilate: Everyone's brain is different, so how can we know that the blue I see is the blue everyone else sees?

The room goes dark. No one can see your face turn pale; color is a function of light.

People tend not to like things that are so damned slippery, so we try our best to make sense of it, and one way we do that is by putting words to the sensation we're having right now. We use language all the time to help us make sense of the world: It's one of those things that's proved useful. This large furry mammal with the roar and the stripes and the long teeth that will kill me—that I am going to call tiger. The smaller version of that furry mammal, the one that will live in my home and knock things off my shelves while it stares me straight in the face—that one I am going to call kitty. When I use the word "tiger," you will know to run away; when I use the word "kitty," you will know to run toward your shelves to save your knickknacks from gravity-induced destruction.

But language, like color, is also more complicated than we think. Look to your left, where there is suddenly a raccoon. You are startled; you holler, "Oh my God, a raccoon!" The early twentieth-century scholars C. K. Ogden and I. A. Richards pop up from wherever the raccoon came from, freeze-frame this episode, and dissect it into what they call the "triangle of meaning," a psycholinguistic series of interactions among three things:

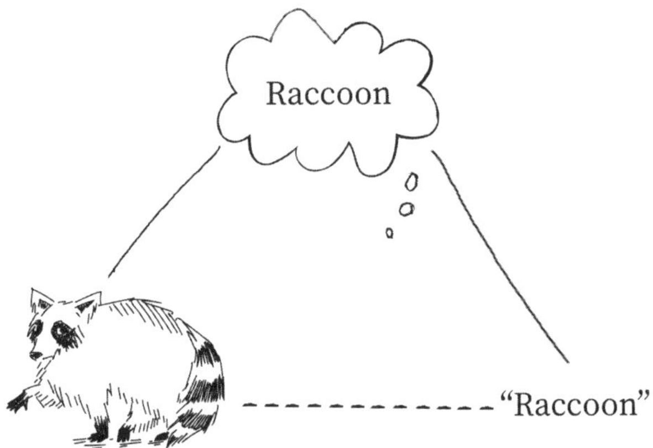

A raccoon walks into a ~~bar~~ book and is trapped in C. K. Ogden and I. A. Richards's triangle of meaning.

There's the referent; that's the raccoon. There's the thought; that's the recognition that the thing you just saw was a raccoon. And there's the symbol; that's the word "raccoon." All three of these things exist in relation to each other, kind of like the points of a triangle, but their relationship isn't equal. There's a solid line between the thought and the referent: The recognition that the thing you saw is a raccoon is caused by that thing actually being a raccoon. And there's a solid line between the thought and the symbol: When we think of raccoons, we automatically attach the word "raccoon" to the thought. But there's a dotted line between the referent and the symbol, because we recognize that the word "raccoon" only *represents* the critter digging through your trash with its freaky little hands. I could use other symbols for that same referent—the German *Waschbär,* the Spanish *mapache,* the taxonomic name *Procyon lotor,* the ASL sign for "raccoon." The point is that whatever I use, the word or sign itself is *not* the actual raccoon, which has left our frame of reference so it no longer has to put up with you shrieking at it or me blathering on about the philosophy of meaning and observation.

But the triangle of meaning gets wobbly when we apply it to color. Look out over a river and unhook your brain from the word "river" for a second. Just look at this one spot that is *not water, you are not looking at water,* and tell me what colors you see: yellow and gray and green and white. Now tell me what color the river is, and you will say "blue," because water is canonically, unswervingly blue. But *why*? The river is blue because water *is* blue. If you have survived high school science, you may start burbling about wavelengths and reflectance and say the color of the sky has something to do with it, I think? Like the rainbow, there's *physical stuff* that happens to make water blue. But why, I ask, is a glass of water clear and not blue? And then we both stare back out, moderately dissatisfied, at this big muddy streak of water that is definitely not blue.

Color is so much a part of our experience that we've drawn over the dotted line of Ogden-Richards's triangle of meaning. Water *is* blue; oranges *are* orange; a tree's leaves *are* green until they *are* red. Color is intrinsic to the thing we're looking at. Plato himself told us that the only two things that an object truly has are form and color, and we all know how smart Plato was! I know that orange *is* orange as surely as its *name* is orange.

Combine these two things together—the hardwired connection between a color sensation and the name we give to that sensation, and the complexities of understanding how the external world, our bodies, and our brains somehow produce this sensation we call blue—and you get a lot of weird claims about color. Did you know, says the person who has read too much internet, that the ancient Greeks couldn't see blue, because they had no word for blue? Did you know that we didn't see orange until the fifteenth century when the word "orange" came into English? Have you heard about the Himba people of Namibia, who can't distinguish between blue and green, because they have only four words for all color? And even as you make a confused face and get ready to make a noncommittal noise of disagreement, you stop. All of these claims seem ridiculous, but it's *color*. Color is odd. You were the person who thought the dress was black and blue, after all.

Everything we currently know about color—and every way that we interact with color, though we never think of it as such—could plausibly be traced back to one point in modern history, which is where our story begins.

Look to your left, where there is suddenly a war.

Dyeing Kills

Color occurs naturally, but most of the color in our modern world is manufactured. Margarine makers, for instance, didn't just forage prepackaged yellow dye from some mountain meadow. Even if the raw materials for that dye were from a meadow, the colorant itself, in its liquid gold form, was carefully, lovingly synthesized.

The production of colorants—dyes, in particular—was a global concern. German dye manufacturers reigned supreme prior to the start of World War I. Read any history of the development of synthetic dyes, and apart from a Brit and a handful of French chemists the pages will be lousy with Germans: unsurprising considering that modern chemistry was essentially born in the nineteenth-century German academy.[1] Synthetic dyes were more colorfast, lightfast, and consistent than natural dyes, and the industry boomed. By 1914, 90 percent of the world's synthetic dyes were produced by German companies.[2]

Then two shots straight at the heart of the Austro-Hungarian Empire, then a local skirmish or two, and then cataclysm. The once thriving dyestuff industry in Germany found itself in increasingly dire straits as the war progressed. Important chemicals needed for the manufacture of dyes were requisitioned for the creation of pharmaceuticals, and the blockade enforced by Britain threatened the ability to deliver one of Germany's main commercial exports to neutral nations. The industry was at risk of collapsing—until the physical chemist Fritz Haber, at the behest of the German Supreme Command, managed to take a common by-product of dye manufacture and weaponize it. Fashion, but make it deadly: In 1915, chlorine gas became the first lethal chemical weapon used on the battlefield in World War I, and the German dye industry was the main supplier.[3]

Dyestuff creation may sound very Ren faire, or like something the Weird Sisters from *Macbeth* would do in their downtime, but modern synthetic dye manufacture is hardcore chemistry with hardcore chemicals. One early synthetic dye, 2,4,6-trinitrotoluene, yielded a lovely yellow color; we know it today by its shorthand, TNT.[4] Many of the chemical catalysts and intermediates that were used to create commercially popular dyes like sulfur black and crystal violet also made great explosives, as was clear from the conflagrations that would break out with some regularity at dye works. Germany repurposed dye works into munitions plants with very little effort, and though Britain did its level best to choke off Germany's importation of some of the chemicals used in dyestuffs and explosives, Haber's discovery was the nail in the coffin. Dyeing killed.

Over in the still-kind-of-neutral United States, the loss of imported German dyes was a huge economic shock. Dyes, and especially those formerly easily imported high-quality synthetic dyes, were used in everything, from money to fabric, house paint to buttons. When war broke out and it was clear that there'd be no more German dyes for import, America scrambled and bought German dye reserves from China and India, but by 1915 they were gone. It wasn't just a fashion crisis: One newspaper writer noted in a 1916 column that the dearth of commercial colors affected so many industries that it could cost the American economy $5 billion, or about $1.4 trillion in modern moolah.[5] It was so severe, it even got its own sound-bitable name: the Dye Famine. There was a push to use American dyes and colorants to stave off a deep economic recession—there wasn't really any other choice—but the cold commercial reality dumped water all over that flag-waving parade. American-made dyes had relied very heavily on chemicals that were synthesized and supplied by the country that had dominated chemistry throughout the nineteenth century: Germany. Without them, the resulting dyes were middling at best,[6] and despite a hardcore propaganda campaign to the contrary, and a call to meet the dye shortage with "a spirit of generous compromise,"[7] people still longed for the vibrancy and colorfastness of the traitorous stuff.

Once Germany reverted to unrestricted submarine warfare in January 1917 and America stopped pretending to be neutral and joined forces with Britain, we had a whole new reason to worry about what Germany's dye works were getting up to. Color, it turns out, was tactical, and just like at home color was *everywhere*. A flare pistol could shoot bright white flares at night to illuminate no-man's-land and enable riflemen to shoot enemies who might be creeping across it under cover of night, but the same bright white flare was invisible during the day, when flare pistols were used for signaling—what color, then, would be dark enough to show against the haze of the front and still be bright enough to use at night? Gray ships were easily seen in profile against the sky and horizon by U-boat periscopes—was there a better color to paint them to help camouflage them from attack? Thousands and thousands of yards of camouflage covers needed to be dyed a consistent color: Make one batch just a smidge too green, for instance, and that camo cover no longer blends in with the mud of the Somme, but stands out like the proverbial green thumb, signaling to German aircraft just where to aim their guns. Aerial surveillance and reconnaissance, both innovations of the war, required photographs taken through high-quality optical glass (Germany's domain) with spectrally sensitive film that could pick up troop movement through smoke and haze: Where was that going to come from now?[8] Even chlorine gas itself became problematic in wartime use—not for its lethality, but because it produced a distinctive yellow-greenish cloud when released that was easily detectable by the enemy.*

Once color became a wartime concern, everything colorful—paints, stains, glass coatings, metal coatings, dyes, enamels—needed to be color coordinated on a national scale. Easier said than done: America lacked Germany's centralized manufacturing structure since it was a tiny bit larger than Deutschland, and now it also lacked the time, money, and raw materials needed to build such an infrastructure. Paints and stains were heavy and expensive to ship over such long distances, so the industry had favored smaller local paint makers instead of big manufacturers. No centralized manufacturing meant that the government relied heavily on already overburdened paint companies to scale up production exponentially and immediately. Very few places made optical or metal coatings, and rarely in the quantity needed by the government. Dyestuffs manufacturers competed with pharmaceutical companies for the same chemicals: Which industry is the War Department going to prioritize when just as many soldiers were dying of disease as were killed in action? The government mobilized the entire economy to support the war effort: Rationing, war bonds, requisitions, and contracts with every Tom, Dick, Harry & Co. became common within months of America entering the war. If large companies balked at letting Uncle Sam decide what they were making and what the government would pay them for it,

* Its replacement, phosgene gas, was much better in this regard: It was colorless. It was also used in the production of industrial dyes.

they might well have found themselves the subject of antitrust suits brought by the government.

When freedom itself was on the line, the name of the game was efficiency, and anything that gummed up the works was a national defense crisis. It was suddenly a liability that America didn't have a centralized and standardized way to describe and create colors. One supplier of dyes for army uniforms, for example, might be running low for any number of reasons—not enough labor, not enough chemical intermediates, sudden plant explosion at the dye works. The government could go to another supplier with a swatch of the dyed fabric they wanted and say that it was olive drab. But that other supplier would have to try to reverse engineer that particular dye—formulas for dyes and colorants were, in spite of being in service to the war effort, still proprietary—and backward engineering color from a finished product is a crapshoot. You have to start by figuring out what color this olive drab actually *is*.

Colorimeters, or instruments to accurately measure color, had been developed prior to the war, but the ones that were easiest to use were the most limited,* and the ones that were the most accurate were the fiddliest. Operating these colorimeters required a deft hand, a trained eye, the right light source, reliable equipment—most of which were in short supply during the war. But even a colorimetric measurement of a color doesn't tell you what chemicals made up the dye that gave you that color. That relied on the chemist's experience on the bench and off: *This* sort of chemical synthesis would probably give you a green close to what you're looking for, but not on short fibers like cotton, unless we tried to use *this* sort of intermediate chemical, though in *this* chemical pro-

* The Lovibond Tintometer was one of these easy-to-use colorimeters. It was developed by a failed gold rusher turned brewer, and it helped measure the color and quality of liquids. Some liquids, anyway. Beer, specifically. Liquid gold—and the Lovibond Tintometer would tell you exactly which shade of liquid gold was the best.

cess *that* intermediate might blow up the plant while we're making the dye, in which case we may want to consider this *other-other* chemical process that would . . . ad infinitum. And once you had a working formula for the color mixture, then what? You still had to synthesize it in a lab; you still had to source the chemicals to make your dyes at scale; you still needed to test how the dye struck the fabric, how it weathered, whether it was going to cause a reaction with the skin of the wearer. This couldn't be done in a matter of weeks. And this is assuming that the swatch handed to the manufacturer was, in fact, U.S. Army olive drab: Different branches of the military used the color names "khaki" and "olive drab" to refer to different colors throughout the war.*

The U.S. government needed to keep tabs on all this, and who better but the science nerds at the National Bureau of Standards? The bureau was officially created in 1901 as the first national physical sciences research lab—a bold move, since America had few working scientists within its borders. (Science was mostly an academic discipline in the nineteenth century; if you wanted to actually *do* the science, Europe was the place to be.) Its mandate was to measure and standardize everything it could, and it was desperately needed: One scientist complained that when dealing with liquid measurements, he had to deal with eight different "standard" gallons,[9] and this in spite of the fact that the U.S. Office of Weights and Measures had been setting standards for, well, weights and measurements since 1836.[10] The early work of the NBS focused primarily on consumer goods, but war changed that as well, and the bureau was quickly called to help fix the issue of national manufacturing unevenness. It had started a running list of all the ways that

* Not just the armed forces, it's worth noting: There are color specifications for everything that the government touched. There was a separate government color specification for the Panama Canal. Not for ships sailing through the canal, or for tagging cargo that went through the canal. No, the color specifications were for the canal itself.

color was weaponized or problematized during the war; it touched on physics, chemistry, optics, illumination, all fields dominated by Germany, all fields that America would need to make serious headway in to guarantee lasting national peace. The head of the NBS went before Congress in 1916 and asked for the wartime equivalent of a unicorn: an extensive increase in funding so they could investigate all these new technologies and shore up American research. "The items submitted—I think I can say all of them—are as fundamentally concerned with both industrial and military preparedness as any that will come before you," he said during the hearings.[11] The budget was approved, which meant that a wartime American Congress threw oodles of money at something that, just ten years earlier, had been considered the province of Parisian couturiers and dress-mad socialites: color standards.

The war was over, but another was just beginning, and butter was the battlefield.

The Yellow Brick Road

In 1869, responding to a call for a cheaper butter substitute for the French armed forces, Hippolyte Mège-Mouriès developed a shelf-stable spread derived from beef tallow that he called "oleomargarine."* Margarine, as it'd eventually come to be called, was vastly cheaper to manufacture than butter, and it was so popular that in the decade following its introduction in the United States in the 1870s, three dozen margarine manufacturers sprang up to meet demand. The American dairy industry promptly had a cow.

One argument that the dairy farmers (and their congressmen) made was that margarine looked so much like butter that the hapless consumer couldn't help but be misled, and this would sink an American economic institution. Margarine is naturally not yellow but white; since it was butter-ish, it was dyed to both tempt the buyer and remind them of the thing it was replacing. Never mind that butter itself was *also* not always yellow, or yellow enough—as any good dairy farmer will tell you, the color of cow's milk and

* Mège-Mouriès created the name from the Latin *oleum,* meaning "oil," and the Greek *margaron,* meaning "pearl": Margaric acid, one component of margarine, has a pearlescent sheen to it. The man who discovered margaric acid was Michel-Eugène Chevreul, a chemist turned dyer whose theories on color influenced scientists and artists for centuries.

cream varies depending on what they're eating—butter was yellow, and yellow was the color of butter. This, at least, was the main assertion under debate during congressional hearings in 1886 on the marketing and sale of margarine. New Hampshire senator Henry Blair (R–Dairy Country) was a particular stickler for the purity of butter yellow / yellow butter: "You may take all the other colors of the rainbow, but let butter have its pre-empted hue."[1] He thought the butter substitute should be dyed a color that wasn't butter colored: red, maybe, or black. Professor Charles F. Chandler of Columbia College and head of the New York Board of Health, who had investigated the production of margarine, tried to convince Blair that some colors were not colors that a butter eater would readily spread on their morning toast. When people want to eat margarine, he said, "they want it yellow." But this was the wrong answer, Blair insisted: If the color of margarine was yellow, then how would people know that what they were buying was not butter? Chandler was unmoved. "It is colored yellow because it is butter, just as much as the other is butter."[*2]

These hearings led to passage of the Oleomargarine Act, which regulated the advertising and sale of margarine (and, of course, taxed it at a higher rate than butter). The act required only that the packaging make it absolutely clear to the consumer that, no matter what color the toast lubricant inside was, it was not butter. States where dairy was an important part of the economy, like New Hampshire, mandated that margarine be dyed—usually a bright pink, but other colors were suggested, like red, brown, blue,

* This exchange devolves into something out of a Marx Brothers routine:

Blair: Is it dairy butter?

Chandler: No, sir; it is not dairy butter; it is artificial butter.

Blair: Please use the two terms so as to distinguish them in your answer. Do you understand that this is a controversy between butter on the one hand and an article which looks to be actually the same thing?

Chandler: Yes; except that it is oleomargarine.

and black. By 1902, thirty states had some sort of color regulation for margarine on the books. Margarine manufacturers fought back: They sold their margarine undyed but included a packet of yellow food coloring so that the housewife could knead it into her margarine, and when the packaging for margarine shifted from wooden crates to cardboard boxes, the cardboard was colored the brightest, creamiest yellow possible.

And consumers responded as Chandler expected them to. The color of money was suddenly yellow. Housewives would cross state lines to buy margarine that was dyed yellow; dairy states began a counteroffensive campaign to remind people, as one full-page ad in *Hoard's Dairyman* magazine hollered, "Yellow Belongs to BUTTER."[3] It turns out that the association we have between an object and its color is so strong that not even an act of Congress can break it.

But things can certainly test it. Bad actors have always played on people's expectations of what color a thing should be to bilk consumers. Pliny the Elder leaned hard into the "Elder" part of his honorific and whined about the charlatans selling inferior pigments as the classical colors of his own antiquity;[*] whole industries sprang up around sourcing and judging which pigments would give exactly the correct shade of ultramarine, madder, Tyrian purple the artist wanted. But these industries themselves were vulnerable to grift and charlatanry. The congressional Butter Skirmish began not with some flag-waving defense of the spectral purity of butter but with an investigation into food adulteration—a huge problem as more and more of our food came from factories some distance away.

* "Nowadays when purple finds its way even onto party-walls and when India contributes the mud of her rivers and the gore of her snakes and elephants, there is no such thing as high-class painting. Everything in fact was superior in the days when resources were scantier." *Salve,* Grandpa Pliny, let's get you back to bed.

Not only did the populace have an inherent sense of what color something *should* be, but they used color as a guide for quality or freshness; oil, cotton, butter, not-butter were all judged based on what color they were compared with what color people thought they should be. This meant that industries that produced large amounts of anything that had a color wanted to find a way to make sure that the color that consumers associated with that thing was the exact color they used in making that thing. Imagine pressing tens of thousands of gallons of grapeseed oil and trying to sell it at market, only to discover that folks thought your batch was rancid because it happened to be just a little yellower than the competition's. The measurement-obsessed National Bureau of Standards was called in to assist. The earliest years of the NBS had been spent on more pressing fiscal concerns—normalizing railroad gauges, standardizing and measuring electrical currents in household items, and pushing for the uniform management of public utilities. But measurement was measurement, and money didn't just flow down rails and through electrical lines, so in 1912 the NBS's Colorimetry Section began developing grading standards based on color for oils (both edible and otherwise) and—yes—both butter and margarine.

Color standards were head-slappingly simple: cards, chips, or liquids that were the "right" color which manufacturers could hold up against their product and squintingly compare them with. They were not common, but they were not exactly new. In the thirteenth and fourteenth centuries, a proto–color standard emerged in medicine. Called (and I am not joking) urine wheels, these images appeared in manuscripts on health, where they were an important and common diagnostic tool. They were depictions of glass flasks filled with fluids of various colors arranged in a circle, from yellow to pink to green to black; each flask would have a description next to it, usually of the color or range of colors that image was meant to represent, and inside the manuscript itself there was usually a little key that would provide a possible diagnosis. A patient would come in complaining of some sort of problem, and the good doctor would ask the patient for a urine sample and pull out a urine wheel. By comparing the color of a patient's urine with the diagnostic criteria on their urine wheel, a physician could tell you that orange urine meant you needed to drink more water, pink urine meant lay off the beets, and black urine meant your liver was diseased and you could probably use a good bleeding.*

Urine wheels weren't a proper color standard, because the colors used to create them weren't themselves standardized. Every urine wheel's samples were slightly different. Paints in the thirteenth century were all handmade, and two artists using the same pigments and the same binders could end up with different colors depending on the preparation of the pigment, the fineness of the pigment's grind, the proportion of binder or suspension to pigment, the type of brush used, the thing painted upon, and whether the illuminator was pulling a late night and so painting by (dim, yellow) candlelight or decided to take care of it in the (bright, white) morning. Nonetheless, artists throughout the late Middle Ages and the Renaissance still produced color charts with the pigments they had available so that others could both identify the color of things around them and try to represent those things. Color standards became standardized once paints and dyes could be reliably mass-produced in the nineteenth century. This happened to coincide with a big boom in the production of consumer goods, like margarine—which is how the NBS got pulled into the grading business to begin with.

* This diagnostic process is called uroscopy, and it's a field of medicine that still exists today. So every time you go to the doctor and have to give a urine sample in a clear cup, you can thank the thirteenth-century French royal physician Gilles de Corbeil and his treatise *De urinis* (On urines).

So how did the NBS decide which color was the ideal color, the color de la color? Scientifically, of course. No eyeballing here—they measured. Using the technology available to them—everything from Maxwell spinning discs, which looked like a multicolor mechanical top on its side that, when spun at high speeds, would cause your eyes and brain to visually blend the colors on that top into one composite color, which could then be matched to a sample, to newer photometers, which measured the strength and wavelength of light reflected from a surface—they studied and recorded the minute variations in color for the thing to be measured, from the highest-quality stuff to samples that were just above garbage. They narrowed the range of "acceptable" color variations from toward the top of the quality scale until they had an ideal color—the average of all the best-quality goods they were measuring. Then they printed those standards under exacting color tolerances and sold them to industry graders (and, for those with moolah, producers). And they were popular. A color standard was the difference between a flush year and bankruptcy.

Money talks. All sorts of industries began to inquire about all sorts of color standards. Can you do one for evaporated milk? Egg yolks? Red meat? What about teeth, can you give us a color standard for dentures?* Unfortunately, the staffing at the NBS and its sister group, the USDA, couldn't keep up with color standards and their other work. The color problem was becoming something that couldn't be dealt with haphazardly, industry by industry, client by client. Even before America embroiled itself in World War I, the scientists at the NBS realized that a system to beat all systems was needed.

This idea also wasn't new: Philosophers, gentlemen-scholars, and nineteenth- and twentieth-century scientists had all taken a

* There is a surprising—maybe even distressing—amount of literature on color standards for teeth.

whack at such a transdisciplinary system. Patrick Syme, a Scottish artist and art teacher best known for his botanical paintings, took an earlier naturalist's (all prose) system for categorizing minerals by color and translated that across the natural world—using color swatches that he not only named and described by their pigment makeup but also keyed to an analog in the animal, vegetable, and mineral kingdoms. Duck green, then, is "composed of emerald green, with a little indigo blue, much gamboge yellow, and a very little carmine red," according to Syme; it appears in the neck of the mallard, in the upper disk of yew leaves, and in the mineral ceylonite.[4]

Syme's book *Werner's Nomenclature of Colours* (named for Abraham Gottlob Werner, the original mineralogist to devise this color ordering system) was a big hit with naturalists. Darwin had it with him aboard the HMS *Beagle* and used it to describe the colors of the various flora and fauna he ran across; John Richardson used it to describe the zoological specimens collected by the British in colonial America and accurately paint them in *The Zoology of the Northern Parts of British America;* and Andrew Smith referred to it when compiling his *Illustrations of the Zoology of South Africa*. It was a commonsense approach to cataloging the 108 colors contained therein and sparked a whole wave of color dictionaries and color standards. Syme was great, but only 108 colors? Everyone interested in color set out to outdo one another. One dictionary of color, written in 1912 by Robert Ridgway, the curator of the Department of Birds at the United States National Museum, clocked in at 53 color plates and 1,115 named color samples. That's not to say that Ridgway named 1,115 colors for his dictionary: He was, like other dictionary writers, just compiling commonly used names for the reader.*

* In the prologue to his *Color Standards and Color Nomenclature,* Ridgway lists at least nine other color dictionaries or standards as sources for his colors, as well as

It would be easy to think that this is all ego—*mine is better*—or greed—*mine is cheaper*—but to someone with a lexicographical bent there is something compelling about this sort of hyper-minute categorization. Picture a hummingbird mid-flight. Now zoom in to the ruby blaze at its throat—iridescent, shining, purple in one light, bright crimson in another, a mud gray until the flit into the sun, each feather standing out like scales against the bright coin of its neighbor. Keep going, keep going: The end of each ruby feather is pleated like a crinoline, each pleat made up of chevrons constructed from nanostructures that capture and refract light like a soap bubble in the sun, each chevron ranging from fiery orange to lemon yellow to plummy magenta to true ruby. Pull out the microscope and go deeper in. You will see the nanostructures in question swirling inside each of the feather's barbules and be lost in a thousand cathedral windows of atomic color: sapphire blue and sky, emerald green, Kelly green, grass green, green green, Saturn orange and khaki, flecked gold and turquoise, sunset red and blood red, the brightest magenta and the palest leaf green kissing at the very tip of one barbule of one barb of one vane on a single feather that is, in its entirety, the size of a grain of rice. Now multiply that across all the flora, fauna, *mineralia, atmosphera* of our universe: all those billions of individual, observable colors jumbled together in infinite variation. To see just one example is to revel in an entire world; to be so immersed in it that one can delineate and capture each subtle shift in hue is an act of devotion. It is revelation in its true etymological sense: a lifting away of the veil, an invitation into mystery. How can any true investigator keep such wonder to themselves?

---

the colored papers used in education and made by Milton Bradley (yes, *that* Milton Bradley) and Prang, as many commercially produced artist's oils, watercolors, and dry colors as he could get his hands on, coal-tar and aniline dye samples from four companies, chromolithographic inks, embroidery silks, "etc., etc.," he finishes.

The downside, of course, is that color is all around us. There's a lot to catalog, and a lot of people who want to catalog it.

There had been color standards and color description systems developed (kind of) in cooperation with the NBS before. Albert H. Munsell, a painter who studied at the École des Beaux-Arts in Paris between 1885 and 1888 and who later taught art at the Normal Art School in Boston, became a somewhat surprising partner with the NBS. He was fascinated by color—the swiftly emerging science of it, the natural patterns of it in nature, and the orderly presentation and description of it for the artist so they could more efficiently lay out their palette. As an artist and teacher, he wanted something precise for his students, something that encompassed the best of the fine arts that was also grounded in the latest scientific discoveries about color but was keyed to the student. Munsell's goal: Train up the youngest generations to see color scientifically, and then you'll have no trouble standardizing all discussion, matching, and creation of color across science, art, and industry. Music of all kinds could be adequately broken down and communicated in a systematized structure of notes, rests, keys, time signatures, he noted. Why couldn't color?

The way we see color is the foundation of his system. "The eye judges by sensations and is the ultimate test," Munsell wrote in his diary in 1900.[5] Every color in his system would be separated by "one step" of visual perception. Hold up two cards of identical color—let's say red—to a viewer, and then begin swapping one of those red cards with a card whose red is adulterated, drop by drop, with yellow. When does the viewer say, "Oh, those don't match"? That is one single step of visual perception. If you took the scientific photometric measurements of each color and then mapped them according to this step of perception, then teaching the relationships between all colors would be hopscotch easy. It would also mean that you could easily notate those colors.

In his 1905 book, *A Color Notation,* Munsell put forth his initial plan for this descriptive color system. Imagine a peeled orange. Now un-orange it into segments: Each segment of that orange (the fruit) is one of the colors of the rainbow. Red shifts into orange shifts into yellow shifts into green, all the way around until purple snugs up against our original red segment, making an unbroken sphere of color. The sun shines down on your orange such that the colors on the top of your orange are lighter; the colors on the bottom of your orange, being in shadow, are darker. The colors on the outside of your orange are vibrant; as you peer deeper into the segments, which we are pretending for the moment have become magically transparent, the color of each segment gets less saturated as it moves toward the murky middle. If you squint, you can make out individual juice sacs, the colors of which make up each ur-color, their cell walls just barely visible, evenly spaced one step of perception apart. Congratulations, you are holding in your now sticky hand an approximation of all the colors in the world, plotted out in what scientists call a color space.

Now cradle that sticky orange and gently squash it. As Munsell measured all the paints and pigments he could get his hands on, he found that if he consistently applied the "one step" rule to the colors he was measuring, they didn't fit in a tidy little sphere. There were a lot more yellows perceptible to the average viewer than fit tidily on a sphere; we saw lots and lots of minute shades of blue purples; we were somehow lousy with greens and relatively bereft of reds. But science is science. Munsell diligently plotted them out and came up with this lovely, bananas thing:

The Munsell color solid from the yellow-green-blue side. Illustration courtesy of the Munsell Color Company through Dorothy Nickerson and printed in *Webster's Third*.

The other side of the Munsell color solid, from the blue-purple-red side. Illustration courtesy of the Munsell Color Company through Dorothy Nickerson and printed in *Webster's Third*.

This is a visual, three-dimensional representation of all the colors that we see, evenly spaced apart, mathematically and scientifically sound, even if the output looks like Salvador Dalí's fruit basket. Munsell created a nomenclature for his system based on three measurements—hue (or the color of the wedge you were looking at, represented by a letter abbreviation), value (how light or dark a color was, represented by one number), and chroma (how saturated or unsaturated it was, second number)—that "eliminates the personal bias . . . and is an approach to a standard that can be depicted at will by numbers."[6] Detailed wedge by wedge in the 1915 *Atlas of the Munsell Color System* and then expanded and re-released in 1929 as the *Book of Color* published by the Munsell Color Company (a company founded upon Munsell's death in 1918), it was perhaps the first color ordering system that was scientifically based yet took the average viewer's own color acuity into account, and it was a revelation.

The NBS had been watching Munsell's work carefully. The Colorimetry Section had assisted Munsell with measurements of the color chips he began to assemble for the *Atlas,* tested his own photometer for its accuracy, and discussed the best sort of lighting and illumination sources available for taking color measurements, and was impressed. He was the rare artist and teacher who had a truly scientific interest in color, and anything they could do to nurture a solid color standard was only going to benefit the NBS and, in turn, the country.

The NBS had gotten used to monitoring all public and (if it could) private scientific endeavors since the war. Edward B. Rosa, head of the NBS, noted in a 1921 expenditures report that "our industries and our civilization are largely based on science and its manifold applications"[7]—specifically, science funded and overseen by the government, and the NBS was the queen of governmental scientific applications. Accordingly, NBS staff became valuable assets after the war: Despite the increased funding for the NBS, scientists leaked out into the private sector and with them went NBS innovations. Scientists who worked on color accuracy in military reconnaissance photography went to Eastman Kodak; physicists who developed pigments for signaling glass went to work for Corning Labs or various ceramics manufacturers; artists who were conscripted into service coming up with camouflage schemes for navy ships and machine-gun racks perched atop the trenches went off to teach at art schools or design for fabric weavers; chemists beelined right for DuPont; others went to Sherwin-Williams or the National Lead Company, where they put their know-how about pigment reflectance and colorfastness into paints, enamels, and coatings for everything from houses to cars to radios to toys; plenty went into the printing industry, where their knowledge of pigments in suspension would eventually lead to better four-color printing; the optics guys headed to General Electric, where they worked on refining lights for optimal daylight color, or turned an eye to improving the instrumentation used in color measurement.

If civilization was based on science and its manifold applications, civilization was about to get really colorful.

Out of the Blue

The top floor of the Myrick Building in Springfield, Massachusetts, had quite the view. It was the tallest building in the neighborhood, one of the first skyscrapers in Springfield, so was placed to see and be seen. To the west, the cadet-blue ribbon of the Connecticut River, fringed with sap-green elms, maples, chestnuts, wended its way under the Memorial Bridge; follow it north and see the Kelly green carpet of woods crumpling over Mount Tom; look south and you'll see the brick-red and alabaster buildings of downtown edged with a few boxy mills, and the silver shimmer of the Connecticut peeking through the tree cover. In the fall, the view looked like a postcard out of *Yankee* magazine; in the winter, once the river froze over, it was a Charles Wysocki puzzle. A shame, then, that the top floor of the Myrick Building was occupied by the lexicographers at the G. & C. Merriam Company—men who directed their gazes firmly deskward, or, as was the case with one Paul Carhart in 1930, possibly ceilingward in frustration/supplication/exhaustion. The professional gaze must be up or down, but certainly not out the windows: The view would have been too inviting.

Paul Carhart had joined the Merriam-Webster dictionary company after finishing his PhD at Yale, just in time to work on the supplement to *Webster's International Dictionary* for 1890. He was no ordinary hire: He had been trained in phonology, the newest and most prominent subfield in modern linguistics and the latest academic field that Merriam had identified as critical to the scholarly success of its dictionaries.* Rather than shunting him into a desk and barricading him in with printing proofs, the editor in chief sent Carhart off to study with the two most prominent scholars of pronunciation and phonology around: Henry Sweet in England and Wilhelm Viëtor in Germany. When he returned from the Continent, he was immediately put in charge of all the pronunciations at Merriam-Webster.

Carhart was a dictionary man, through and through. When the president of Merriam died in 1908, he was one of the pallbearers;[1] twenty years later, when one of the publishers of Merriam joined the choir invisible, Carhart served the same role. His work on an unwieldy brick of a book led to him becoming one of the investor-owners in a bookbindery in 1911. In the office, he was described by colleagues as friendly, modest, and a sharp scholar: Management wasted no time in making him one of the company's spokesmen for the value of a scholarly approach to the language, and he was trotted out often to answer press queries or to speak to regional groups about the importance of good English. He might not have exactly loved the public eye: The social notices in the Springfield papers heralded the annual summer removal of Mrs. and Mr. Carhart and family to their farm in Becket, Massachusetts, about forty miles away from the hustle and bustle of Springfield. Internal company memos prove him to be a tidy, thorough editor, and for his pains he was given more pain: In the mid-1920s, as the company

* Merriam had, half a century earlier, been pooh-poohed by scholars for ignoring the newest language science coming out of Germany. It wasn't going to be called out again.

began the ramp-up to the new edition, he was promoted to managing editor. The preface to *Webster's New International Dictionary, Second Edition,* lays out Carhart's role in one terrifying sentence: "Upon him early fell the responsible work of conducting the business arrangements with the special editors and of carrying on with them the necessarily enormous correspondence." If "enormous" doesn't adequately explain the scope of correspondence Carhart had to keep up with, about a dozen pages later in the *Second,* the section that lists each special editor and their qualifications notes that there were exactly 207 special editors.

During the nineteenth century, dictionaries grew fatter, not necessarily on new words, but on information tucked away inside definitions, like history lessons or moral tales, and slapped into the back of the binding, where you'd find essays, charts, and lists. They were relatively expensive books, so if you were going to spend the equivalent of two months' salary on a single volume, it should not only tell you what "escapement" means but give you a complete essay and illustration on its construction and workings—and list the heathen gods of Greece and Rome, and all the incorporated towns in America, and provide sample letters for business, and detail a timeline of all history from the creation of the world and birth of Adam and Eve (4004 BC, in case you're wondering) through the date the manuscript was sent off to the printers (1804).* Dictionaries were now a one-stop shop for everything you needed to know about anything.

Among that "anything" was science and technology. Thanks to the Industrial Revolution, science and technology were booming fields, full of rich vocabulary and new ideas that needed expostu-

* This appeared at the back of Noah Webster's first dictionary, published in 1806 as *A Compendious Dictionary of the English Language.*

lating in the clearest way possible. The preface to one gargantuan dictionary of the late nineteenth century, *The Century Dictionary*, notes that "this technical language is, in numberless instances, too closely interwoven with common speech to be dissevered from it."* If science and technology were part of our everyday lives, then the language we use to talk about those scientific and technical achievements was as well.

The problem, of course, was that most dictionary staffs tended more toward the types of nerds who could corner you at a party and talk for thirty minutes about Chaucer or Molière without requiring your participation at all, as opposed to the types of nerds who would corner you at a party to tell you, in great detail, how your flush toilet works. The answer was to hire out for knowledge, which Merriam began doing with abandon in the 1880s. Scholars and scientists were routinely contracted to review and write the definitions in their particular area of expertise, for which they'd get bragging rights and a nominal fee. But those specialists had to be wrangled, edited, coaxed, prodded. That fell to the in-house staff. Or, rather, to part of the in-house staff—in this case, the part of the in-house staff who sat at Paul Carhart's desk.

One field that had slyly scooched from the humanities into the sciences was color—again, something established by *The Century Dictionary*, which hired Charles Sanders Peirce from the U.S. Office of Weights and Measures to handle weights and measures, and who, in turn, decided that as a scientist and logician he got to define whatever he wanted to and that included color terms.† At

* No matter how helpful a dictionary intends to be, there are still some editors who cannot but express themselves most obfuscatorily.

† Peirce's color definitions appear to be, much like Peirce himself, sui generis. They mention microns, wavelengths, luminosity, chroma, brightness, sensations, and pigments—and sky, hair, blood, vegetation, pure snow, and the absolute dark. The masthead notes Peirce was also responsible for "Metaphysics," in case you couldn't tell.

least Carhart had a pattern to follow: He went to the twentieth-century version of the Office of Standard Weights and Measures and ended up finding, at the end of the rainbow, studiously analyzing it, Irwin Priest.

Priest was at the top of the color game (such as it was in the late 1920s). After getting his bachelor's in physics in 1907, he was snapped right up by the National Bureau of Standards, where he spent the rest of his career. He was a remarkable enough physicist that after only six years at NBS he was made the head of the Colorimetry Section and was given the first honorary lifetime membership in the American Oil Chemists' Society in recognition of his scholarship on oil color grading. It was a good time to be at the NBS: By the time Carhart came knocking, the NBS housed more notable physicists than Harvard, General Electric, Bell Labs, Johns Hopkins, Columbia University, the California Institute of Technology, Yale, and Cornell, and Priest was among the most notable. Priest came with the sorts of bona fides that look good on a dictionary contributor's page, and so Carhart was relieved. Priest himself was eager to see the new science of colorimetry reach a broader audience—and what better didactic tool than *the* Didactic Tool?

The two men were cut from the same cloth, though they sat on opposite selvages of the bolt. Carhart's letters read like eighteenth-century patronage requests: a little rambly, full of adverbs and qualifiers, quietly self-important and obsequious at the same time. He makes free use of the verb "shall" and the royal "we" in laying out exactly what Merriam needs from Priest. Priest's responses, on the other hand, are governmental in the extreme: typed impeccably on plain white paper, with a subject line, each paragraph numbered and spattered with a good amount of scientific jargon that no one outside the Colorimetry Section of the NBS would understand. For Carhart's "we," Priest responds with "I"; in answer to Carhart's dictum that all definitions must "convey some definite and intelligible meaning to the reader," Priest unfurls phrases like "a unified exposition of the fundamental nomenclature of colorimetrics" and "Maxwellian trilinear coordinates."[2]

Both men were consummate professionals, and this is where the fabric strains and begins to fray, because their industries were zooming off in opposite directions. Despite the push into the twentieth century (only thirty years late), reference publishing was still stuck somewhere in the 1860s, when the most important name on the book was the one on the cover—Webster, Worcester, Century. If a specialist was fortunate enough to be approached by one of the reference juggernauts, the appropriate response was to throw yourself under the erudite wheels of lexicography and be anonymously crushed. Carhart assumed that he would do with Priest as he had done with the other 206 special editors: Give him a list of words to define and a copy of the previous printing of the unabridged, and collect the fruits of Priest's diligent yet directed labor.

Priest, however, was one of the top scientists in a booming discipline that touched on nearly every other scientific field imaginable. Physics, chemistry, optics, biology, microscopy, metallurgy, astronomy—color was integrally woven into all of them, which meant that the color scientist was competing with physicists, chemists, opticists, biologists, microscopists, metallurgists, astronomers for precious paper spots in scientific journals or science symposia. When Priest began his career, science was a rare occupation. But the government had thrown buckets of money at the sciences after the war, and people went where the money was. Collaboration, yes; cooperation, of course. But before all that, an understanding of who the senior scientist was (him), and who was leader of this outfit (ditto).

And so it quickly came to pass—within just one month, in fact—that Carhart realized Priest was not going to quietly beaver away on the color terms that Carhart had thought he contracted Priest to write. Carhart had already told one of the Merriam secretaries to start cutting all the color-related words out of a previous edition of the *New International* and pasting them onto index cards for Priest's use when Priest sent in a bullet-pointed letter putting Carhart, a *mere editor,* in his place. Priest would begin work on the vocabulary of colorimetry—the science of measuring and quantifying the various attributes of color—and he would write a lengthy article for this new *Second New International* explaining the latest in color science to the dictionary reader. This article would be a signed contribution and approved in advance by the director of the NBS Editorial Committee. It would be accompanied by art and illustrations solicited and approved only by Priest, who also gets complete editorial control over which words he decides to define and which he doesn't. The initial list of words he's working on doesn't exactly scream "color" to the layperson: "spectral distribution," "lightness," "chroma." Priest acknowledges that such scientifically oriented terms are new territory for a general dictionary, "which (as I learned in my youth) is designed merely to record usage," he sneers. He is willing to bring Grandfather Merriam into the twentieth century, but he must be given a free hand. "Most of the technical definitions which I would write could claim but very little warrant in usage," he warns/brags, "and some of them would have no authority except my own personal fiat."[3]

There was one more thing. Priest insisted that he was not going to define the names of colors. "These weird and fanciful names are numbered as the sands of the sea," Priest tersely breaks it to Carhart, "and I would regard their definition as not only a futile waste of labor but impossible."[4] He is a scientist; color names are beneath the scientist.

Carhart was beside himself. The idea of ceding complete editorial control over any part of the dictionary to *a consultant* was anathema. Had Priest never opened a dictionary before? No contributor gets a mention apart from the industry-standard small-caps mention in the preface; not even the in-house editors get to sign their definitions, no matter how good they may be. Priest was work for hire; he was acting like the Sun God. Not to mention the reprehensible notion that a *government agency* would have to approve this article's publication. This was America: Publishers may solicit endorsements from government agencies for the purpose of selling dictionaries, but they certainly don't let those agencies tell them how to write their books.

But what other choice did Carhart have than to make some noncommittal noises and vague gestures of maybe-agreement and slowly nudge Priest in the direction this big book needed to go? He had two hundred of these egos to massage, tens of thousands of entries to wheedle out of these petty dictators of their tiny academic islands. He had solicited the best color scientist in the nation by going to the only place he knew to go. It was 1929, and the *Second* was well under way. He had no time, no connections, and no other resources. He would ignore the comments about the color names—no sense in chasing his expert away—but he did have to draw the line somewhere. He wrote back and gently let Priest know that, no, there would be no signed contributions. Priest responded by not responding for three full months.

So much for professionalism.

You likely share Carhart's confusion over Priest's disdain for color names. The plain facts of English usage are that, by sheer volume, our main vocabularic interaction with color is through color names, not terms of color science or optics or chemistry. How often do you go into a paint store and ask for them to mix up samples by handing them a sheet of paper with trichromatic measurements and diffuse spectral reflectance ratios? "Never" is the correct answer, though I will also accept "what are you *talking* about?" No: Just as Priest and his nerds have their own vocabulary of color, we, too, have a general vocabulary of color. We tell the person behind the counter that we're looking for neutrals (which are not colorimetrically neutral at all), or jewel tones (which include the colors of only some jewels and not others, like diamonds), or pastels (colors named after a type of fine-art crayon that can and does come in colors that are not pastel, sense 2); warm colors, cool colors, strong colors, soft colors. Or we go straight to the wall of paint chips and begin to collect a poker hand of Dark Evergreen, Muted Graphite, Negroni, Plantain Chip, Firecracker, Swiss Coffee, Secret.

Priest's whine that color names are "weird and fanciful" is only half true. Many common color names do have a logic to them, but it's a logic built on human experience and not some scientific nomenclature. There are four main types of color names. We've got the basic color categories—the first colors that we learn as toddlers, the main color buckets into which all other colors get dumped. Generally speaking, that's red, orange, yellow, green, blue, purple, brown, pink, gray, white, and black.* Beyond the basic color terms, there are color names that some scholars call intrinsic. These are color names based on something in real life—plants, foods, gems, animals, earth. "Lime" is (supposedly) the color of limes; "daffodil" is the color of daffodil flowers; "burnt umber" is the color of a type of earth from Umbria (baked), and "raw sienna" is the color of a type of earth from Siena (raw); "cardinal" is the

* These "simple" color names are anything but; see the chapter "The Silver Bullet" for information that will amaze, confound, and possibly anger you.

red of the bird; "mummy" is the brown of the pyramid denizen.* "Based on" is loosely used here: Rarely do intrinsic color names exactly match every specimen of the thing they take their name from. Limes (neither rind nor flesh) are "lime"; daffodils span a range of yellows, whites, oranges, but only one of those colors is "daffodil." Still, intrinsic color names are amazing: a testament to the observational powers of humankind, universal enough that the vast majority of a language's speakers have a general idea of what broad color category an intrinsic color name sits inside, but specific to individual groups, places, times. Paul Green-Armytage, a color researcher from Australia, ran an experiment where he gave a number of English speakers (mostly students) from Australia, England, and the United States five minutes to write down all the single-word color names they could think of (so "blue" was fine, but "baby blue" was not). He got about 270 responses, tallied them, and ended up with a list of about 203 color names, which he sorted by different variables. You can cross-section an entire language's speakers depending on their color names. Green-Armytage found, for instance, that only Americans listed "cranberry" as a color name, and only Australians listed "mulberry." Your field of study or work plays a part: Architecture students named colors derived from built environments and materials (copper, granite, lead, oak); design students listed "magenta" and "cyan," two colors used in printing; painters named "ocher," "sienna," "umber," "vermilion," all of which are the names of fine-art colors or pigments; the members of the West Australian Quilters Association named colors that came from fashion and textiles, like "beige," "cerise," "burgundy," "aubergine" (or "eggplant," if you're a member of an American quilters' guild).

* Literally: "Mummy brown" was a popular color in the nineteenth century, and the pigment was sourced from pulverized mummies. Pigments are weird; Victorians are weirder.

Then we have associative color names. These are names that are usually tied to a person or a place, but there isn't a one-to-one mental association between the thing named and the color. The Alice of "Alice blue," the Althaea of "rose d'Althaea," the Josephine of, well, "josephine"—which Alice/Althaea/Josephine comes to mind, and then, what color do you associate with them? "Josephine" evokes windswept romance, jackets and short pants, France—is it blue? Is it pink? Is it a gray? Or pick a basic color term and travel the world: Chinese blue, Prussian blue, Armenian blue, Olympian blue, Pompeiian blue, Milori blue, Venetian blue, up through Neuwied blue and Bremen blue, then over to Holland blue (but more specifically Leyden blue) and Antwerp blue, then down the coast through Brittany blue to Paris blue, hop the Channel for Oxford, Derby, Victoria, Cambridge, and Eton blues. How are all these blues different? Surprise: Some of them are *greens*.

And then we have the truly fanciful color names. These are color names that are meant to evoke a feeling, and they are highly contextual. I have, in my possession, a sales pamphlet for office cubicle walls—you know the ones, the standard-issue half walls that are covered with a burlap-esque plain-weave fabric and edged with aluminum binding—that gives color samples for all the fabrics you can get your office-drone-box covered in. The colors are all named things like Hush, Mute, Ambiance, Mellow, Muffle, Mystery, Secret, Placid. Are you asleep yet? How's your resting heart rate? You are completely relaxed; when I snap my fingers, you will wake up, and you will increase productivity by 200 percent. If I ask you what colors these actually are, you will use the context in which you've encountered these names (office, cubicle, corporate, blah) and then draw from your various internal associations and feelings about these words. Cubicles are always gray, or beige, or taupe, or another neutral barely-there color, so you'll pull up a mental range of tans and grays and start matching words based on that.

But those associations sometimes steer us wrong. "Mellow," for instance, is the label on a bright, kindergartner-scream orange—certainly not a color I associate with the word "mellow." "Secret" is practically electric blue, as if "practically" would make it a color that you'd put on an office wall. But which employer is going to buy cubicles in a color called Electric? All the fanciful names rely on a complex, individualized interaction between an object, a word, and a viewer—which makes them practically impossible to define in a general dictionary.

Not that dictionaries have done a great job even with the basic color categories or the intrinsic color names. Most monolingual English dictionaries didn't even bother with most colors—even the dictionaries that are supposed to be exhaustive. The color names that tended to be defined were the ones used in literature—heraldry, or art criticism, or philosophy—and most often only if they were tied to a particular pigment or colorant. Samuel Johnson's 1755 dictionary, considered a cornerstone of modern lexicography, enters the fruit "orange" but not the color—though the word had been used to describe one of the core colors since the sixteenth century. He defines "pink" as "a colour used by painters."[5] Which color, Sam? You know, one of the ones used by painters. Even the big dictionaripedias of the nineteenth century tended to enter color names only incidentally—mostly out of encyclopedic or etymological interest. *The Century Dictionary* notes that "Turkey red" is a color from a particular dyestuff (madder) and is really only of interest because it's so named from being first produced in the Levant. More common color names were Johnsonian in their vagueness. "Puce" in the *Century* was defined as "of a flea-color," which is only helpful if you have excellent vision and fleas.[6]

But this sort of imprecision was not going to work if the *Second* was to be a scientifically informed dictionary. If Priest continued in his obstinacy to refuse to even look at the color names, how was Merriam supposed to rise above?

Carhart had no answer. It was Carhart's job to have an answer.

While Carhart fretted, Priest's stew simmered down, and his petulant little stomp-fest eventually ended. Priest's first substantial letter to the office at the end of July 1930 notes that he has not written because "(1) I have had no time to write. (2) I have had nothing definite to say."[7] But now he had plenty to say. He's been working, and he is sending along for review his definition of "color" and "light" (two colorimetric terms). The definition for "color" is "Colors are the characteristics of the several discrete areas of the field of view by which these areas are distinguished one from another," which is vague enough that it could also be used as the definition for "shape," "volume," "variation," or "hairstyle," because all these things are visual sensations used to distinguish one thing from another. The definition for "light" has five different meanings, most of which are not actually meanings but textbook articles, the entry ending with a note at the bottom that reads, "Strictly speaking 1, 2, 3, and 4 are not definitions." He's not wrong.

However, he will deign to discuss the so-called core vocabulary—color names—that Carhart insisted on wringing his hands over. He has a solution for Merriam. McGraw-Hill had just published *A Dictionary of Color,* written by Aloys Maerz, the director of the American Color Research Lab (who, in his spare time, invented a collapsible stereoscope), and M. Rea Paul, a colorist working in the National Lead Company's Research Lab, and if Merriam wanted the most scientifically grounded and up-to-date information on something as specious as *color names,* it had to start there.

Maerz and Paul's bona fides were impressive. A 1938 register noted that Maerz's lab focused on "color measurement, specification, and standardization; color vision; color photography; problems in connection with dyes and pigments; identification of older and ancient substances; historical research; problems of color in industry and sales psychology; [and] consulting services." This meant that on any given day Maerz or his staff could talk with tomato canners, archaeologists, military photographers, art conservationists, salesmen from Woolworth's, silk dyers supplying couture fashion houses, car lacquer chemists, scientists working on incandescent lightbulbs, gemologists, leather tanners, butchers (for grading the color of meat), bakers (for determining the quality of flour by color), and candlestick makers (for suggesting pigments that could stay suspended in wax yet not precipitate out and stain a fine tablecloth). Paul's work at the National Lead Company's lab was more limited in scope—National Lead mainly dealt with paints and other applied pigments at that point in time—but paint and pigments got slopped onto and baked into nearly everything, from children's toys to pharmaceuticals to houses. And one of the things that Maerz and Paul agreed on was that the consistency between the name of a color and the color itself, when multiplied by hundreds of applications, was nonexistent. And no matter how carefully pulled together, no matter how (supposedly) scientifically rigorous, no matter how comprehensive in scope, all the color dictionaries and standards they had access to were not careful, rigorous, or comprehensive enough. So they came up with their own, and their focus, as scientist-businessmen, was providing some sort of order to commercial color names. The preface to their resulting book, *A Dictionary of Color,* begins,

> It has frequently been observed that while standardization has been arrived at in practically all other fields, in the use of color names for identifying color sensations, a condition prevails that is usually characterized as chaotic; a state of affairs conducive

> not only of misunderstanding and argumentation, but even, on occasion, of financial loss. . . . Since the use of color in its brightest aspects has grown common in every day life, in dress, printing, interior decoration, and the growing tendency to use color for exterior architectural embellishment, the need has become urgent for standardization and for a complete reference source of color names.[8]

A Dictionary of Color is a tremendous work: It's still considered an authoritative color standard today. The color plates are, of course, the ooh-and-aah parts of the book, more than seven thousand color chips arranged by hue and value, but it's the front and back matter that make the book worth its weight in gold. The preface includes an extensive breakdown of the types of color names in use, how they interface with other color names, how colors are described by various other standards, how Maerz and Paul went about measuring the colors included in their dictionary, what color is and isn't. There's an appendix that lists all the ways that literature on color describes the three main attributes of color (which they call "hue," "purity," and "value"); there's a multilingual appendix for color names; there's a list of how often a particular color name is used among paint manufacturers and textile standards; there's a bibliography of sources that goes back to 32 BC. The back matter includes a history of color standardization, historical notes on major color names, and an index of the thousands of names covered by the *Dictionary*, including a key telling the reader which field or source that color name is from. The explanatory information makes up more than half the book.

It's the best scholarship available, Priest notes. He suggests that an in-house editor go through Maerz and Paul, pull color names they wanted to enter, and then create a representative committee of color professionals to evaluate the Maerz and Paul list for entry. Priest is not about to be part of that stupid committee. "The value, currency, propriety or good or bad usage of popular, artistic, poetic, literary, fantastic, bizarre, fashionable, and 'sales-value' names of colors" is *not* his interest or specialty, he sniffs.[*9]

One more thing: Deadlines upset his "equanimity and continuity of thought," so he's going to ignore them and advises Carhart never to bring them up again.[10]

You can practically hear the groan Carhart likely made upon reading that—the Merriam-Webster editorial floor is full of these spectral groans, which cling to the living like dour jellyfish. Dictionaries only ever make it to market because a production editor is daily breathing down every editor's neck, muttering the deadline over and over again in a not-quite-threatening undertone, until everyone on that floor can feel the weight of the deadline in their own bodies, right near the solar plexus. And every lexicographer alive knows that defining by committee, as Priest suggests they do, is the quickest way to create territorial skirmishes and lifelong enemies, not usable definitions. If the established authorities can't agree on a single, consistent definition of "dusty aqua," why should corralling them in an airless conference room stuffed to overflowing with old dictionaries and bad lighting change anything? Consensus had to be found in actual usage; that was the whole damned point of getting a color consultant, who lived and moved and had their being in color and its applications, involved.

The editorial office was closing for summer vacation, so Carhart had to dash off a reply. "Thank you very much for your wonderful letter of July 30," he seethes. "It is the finest things of the sort that we have ever received, to the writer's knowledge."[11] But they were stuck. Priest was the expert; to dump him a month

* Priest mentions sales and salesmen three times in this section of his letter, each time underlined or in what I assume are scare quotes. If he's doing his best to distance himself from anything commercial, it's working.

before they needed the color terms for the letter *A* was madness. Carhart wheedled and buttered and blandished his angry editorial heart out; still, they promptly agreed to everything Priest wanted. Except for the pestering.

After sending two unanswered letters to Priest in August and September, Carhart imposed himself on the expert in person in Washington, D.C., on October 21, 1930. The September deadline for the letter *A* had come and gone, and nothing had arrived from Priest. Things were past "getting dire"; they were already "dire" and sailing well on their way to "catastrophically bad." It had to be extravagantly terrible to pull Carhart—the editor around whom nearly all the production and editorial coordination for this book whirled—away from his family and his office and force him to travel to D.C. on an overnight trip to persuade this mad scientist to comply with the terms of their agreement. And given how penny-pinching the president of the company was, the request for an extra expense just to glower menacingly at a consultant was likely not well received. It appears to be the only such trip that Carhart made; of the two-hundred-odd consultants, only Priest required so much from him. But the production schedule slows for no one, expertise notwithstanding. In layman's terms, fish or cut bait.

No record exists of the conversation, but we have Priest's immediate response to it, dated the next morning. It opens with no preamble, no niceties—not even any spleen that Carhart essentially invited himself to Washington, D.C., to impose himself on and irritate a very important, very busy scientist. Priest, the man in charge, has capitulated: The work Merriam wants from him is beyond him. "Pursuant to your oral request of October 21," his letter begins, "I have written to Dr. Godlove as per copy herewith."[12]

The Golden Boy

Isaac Hahn Godlove, or I.H., as his friends and colleagues knew him, had serious scientific chops but none of the nerdy standoffishness of the modern scientist. He had graduated from Washington University with a bachelor's in chemical engineering in 1914, and immediately followed that with a master's in chemistry from the same institution in 1915. In 1916, after a year of teaching chemistry in Norman, Oklahoma, he was accepted into the PhD program at the University of Illinois Urbana-Champaign.

This was no small feat. The chemistry department at Urbana was under the direction of William A. Noyes, the original chief chemist of the NBS, whose specialty at the bureau was atomic weights and who was regarded as one of the best organic chemists in the country. He had been lured away into academia with the promise of money (which was in scant supply at the bureau prior to its expansion during and right after World War I)* and the charge to build the best graduate department of chemistry he could. And he did: Noyes brought some serious chemists under the Illini banner, including Worth Rodebush (who pioneered work in infrared

* Yes, Virginia, there was a time when academia paid.

absorption methods for studying molecular structures and later discovered rocket propellant), C. S. "Speed" Marvel (who helped establish the field of organic polymers), and Roger Adams (whose sexiest research involved isolating cannabidiol and comparing it with THC*). When Noyes took over the chemistry department in 1907, there were eleven teaching staff and seventeen grad students. Ten years later, with the postwar science boom just revving up, there were twenty-four professors or instructors and seventy-two grad students (forty-three of whom were teaching assistants or research assistants), including Godlove.

Godlove was at Urbana for a total of ten years as both a student and an adjunct. His PhD thesis was titled "Application of Spectrophotometry to Organic Color Reactions," and it was undoubtedly this line of research that eventually led him to the Munsell Research Lab, where he was hired in 1926 to be its research director.

Very few publicly available photos of I. H. Godlove survive from this early period of his life, so best to go to the man himself. Since college, I.H. had been an avid amateur artist, with the emphasis on "amateur": A feature written about him noted that Godlove's father had established a society to support art, "which turned eventually into one for buying meals for indigent artists; and two to six were always cluttering up the dinner table. He decided against art as a career."[1] Godlove drew, and most of his line drawings were copies of pictures or other works of art—the graces, the starlets of the talkies, national figures, portraits of friends. There's only one self-portrait in his portfolio, titled *Myself, by Myself.*[2] It is undated, but it tells you everything you need to know about I.H. His other figure studies all decorously occupy the middle of the page, leav-

* This is research he undertook at the behest of the Narcotics Bureau of Washington; he was later investigated by the FBI under Herbert Hoover and was initially denied security clearance to work on synthetic rubbers during World War II by the same government that had approached him to do the marijuana-related work in the first place.

ing a wide margin for a mat, maybe a frame; in his self-portrait, his head takes up nearly the entire page. The artist doesn't flatter: His ears stick out, he has recorded the laugh lines that frame his mouth, and he faithfully maps his widow's peak, which is racing backward for the crown. Given that, maybe we should forgive him for detailing each eyelash, the fullness of his eyebrows, the tidy bottle brush on his upper lip. Godlove is staring straight ahead; his eyes are clear, and you can see reflected light in each iris. He is smiling slightly, lips parted as if he's just had a funny thought and is mid-inhalation so he can share it. It is not a portrait that tries to convince you, haughtily, that you are looking at the smartest person in the room. It's a picture of someone who is curious, someone who is eager to be everyone's friend.*

That curiosity and friendliness served Godlove well. Though color measurement was very much a science at this point, color as a field of study was not. You can get a sense of this browsing through *Color News,* the short-lived publication of the Munsell Research Lab. Inside its pages were articles on the history of color starting with the evolution of eyespots in denizens of the primordial ooze; color, aesthetics, and design; common ways of measuring color; a short bibliography of color-related books; notes, jokes, and observations about color in society; and a colophon letting you know the Munsell values of the colors used in printing the cover image. Science was pulling color into the twentieth century, but it still had a nineteenth-century gentleman-scholar's air about it. In short, it was a field where wide-ranging interests and wide-eyed curiosity were still welcome.

Godlove had both. He had been deeply involved in the production of the *Munsell Book of Color* and made connections like Priest at the NBS through his work, but left the lab in 1930—no sense in

* It is also, like all of Godlove's drawings, only in black and white. Having studied color, he knew his limits.

staying now that the *Book* was done. From colorimetry to curation: Right after he left, he was asked to oversee, as one of the directors, an exhibition on color for the Museum of the Peaceful Arts in New York City. The exhibition was to be, in the words of one of the academics involved, "the first comprehensive effort yet made to indicate the use and future possibilities of color in virtually all departments of modern life," which would "bring to the public a better understanding of both the scientific and artistic aspects of color."[3] Priest had helped get him the post, and the work was extensive. He was in charge of "planning, development, and direction of exhibition on chemistry, physics, psychology, and industrial phases of color; executive work, securing great number of exhibits from important American industrial organizations; public lectures; broadcasts," according to his CV.[4] This sounds padded out, but by all indications it wasn't: Priest noted in his surrender letter to Carhart that the only way he'd allow Godlove to take on any of the color work for Merriam was if there was ample time left for him to work on the exhibition, which was, as Priest put it, "not a mere routine seven-hour day with plenty of play time." Whatever the agreement between Priest and Godlove was, and however Godlove interpreted it, is unknown: No record of it remains.

Before Godlove had sent in a single definition, the relationship between himself and Merriam was fraught. This seems to have been mostly due to Priest, who made it clear in his letters to Merriam that he would steer young Godlove in the right direction and help him devise a plan and pattern for the definitions, based on the latest research—but entirely on his own schedule. Carhart wrote directly to Godlove at the end of 1930 asking if it might be possible to have a conference with Priest sooner rather than later, and inquiring discreetly about whether Priest's claim that Godlove was far too busy to do anything extra was true; Priest got wind of it and promptly lost it. "I am still planning to give most of December to the actual preparation of this article [for the entry for "color"], BUT I POSITIVELY WILL NOT AND CANNOT GIVE ANY ATTENTION WHATEVER TO THE DEFINING OF COLOR NAMES BEFORE THE DRAFT OF THIS ARTICLE IS FINISHED," he type-hollered to both Godlove and Carhart. A line had been crossed (again): Godlove was merely an underling, and Priest was still the one in charge of all the color work, and that included all the preliminary work for the color names, which were merely going to be scribed out by Godlove (and then only because Merriam was so rude as to demand that someone handle them). Priest was so upset at the pressure that he resigned, "to take effect automatically and immediately upon the receipt of any further communication urging or requesting me to cooperate or give any further advice on the defining of color names before January 1, 1931."[5] If Merriam-Webster wanted the finest minds in color to write their dictionary for them, they needed to stand back and give genius (senior) some breathing room.

Carhart sent back the required blandishments to keep Priest on the job and gave both him and Godlove the required month of downtime. But as soon as Priest wrote in January to say that he had spent December on the color article, and sadly, the muse had not struck to his satisfaction so it would be some time yet before he could do anything else, Carhart dropped the genteel nineteenth-century scholar shtick. They were already overdue for getting everything in letters *A* and *B* so they could send those letters off to the compositor; Priest had not sent along any of the seven hundred color-related terms he had promised, let alone that damned "color" article; there were still hundreds of color-name definitions that needed to be revised, all of which were banded together and sitting on Carhart's desk; the production managers were making more frequent trips around the editorial floor, brows and shoulders increasingly knit. Carhart bypassed Priest and wrote straight to Godlove once more in mid-January 1931. "We have been casting about in our minds for a plan to bring together all the remnants of our former plans that were worth saving," he says, "and to make one last big effort to get the Colorimetry and Color Names started on a satisfactory basis."[6] If Godlove can't get started immediately, Carhart warns, then Merriam will be forced to skip the color names entirely and just reprint old material.

It got the job done. Thirty-year-old definitions for the terms in a field that was growing by leaps and bounds by the day, in a dictionary that was supposed to be *the* most authoritative, *the* most up to date, *the* Dictionary? Never! Without asking for any assistance or guidance from Merriam or Priest, and in spite of working fourteen-hour days as the color-exhibit curator, Godlove got right to business. He decided, like Priest, that the color definitions needed to be drawn from some sort of authority; he decided to work from Maerz and Paul, per Priest's suggestion (though he ignored Priest's recommendation that a committee decide which color names to enter), and to use Munsell as the explanatory system for all the Maerz and Paul colors. Munsell, a commonsense bridge between science and general use, would help readers understand the relationship between all these colors; Maerz and Paul, of course, had taken an extensive industry- and science-based catalog of color names worth knowing. All it needed was Priest's approval.

Alas for Godlove. Though Priest was not actually willing to define the color names, he still clamped on to them like a dog worrying a bone. Priest demanded Godlove's work stop and Carhart threw up his hands; he was done. A conference between Godlove, Priest, and Thomas Knott, the general editor for the *Second* and Carhart's more senior partner on the editorial board, was arranged for early March, and it was there that Priest laid down the law. Munsell would *not* be used; Maerz and Paul would *not* be referenced; the definitions *would* include the term that Priest had just coined in place of Munsell's "value" and "chroma," the incredibly clunky and nigh unpronounceable "hueivity."*[7] He dictated the defining formula to be used for each color—"a color, red/yellow/green/blue in hue, of high/medium/low hueivity and high/medium/low brilliance." These decisions don't reflect any important scientific fundamentals; these decisions appear to be made because Priest—not Merriam, and not Godlove—is the boss of color, that is why. Merriam would send Godlove the five hundred or so entries for colors that appeared in the 1925 edition of the *New International Dictionary* for revision, though Priest deplored the idea of entering color names based on actual written use of color names—so uninformed and imprecise.

Godlove maintained that he wouldn't *mention* Maerz and Paul, but he really needed to use some sort of scientifically rigorous survey of colors and color names to define the colors, and Maerz and Paul had the most complete list of current color names around. Using Maerz and Paul as a source immediately increased the color-name coverage by several orders of magnitude, and bigger was better. Priest shuddered, but agreed, and Knott was satisfied with this, which meant that Carhart was (or had to be) satisfied. It might well have been that roping and wrangling a stable of geologists, chemists, physicists, botanists, biologists, and a Whitman's

* Priest didn't particularly care that the word was new; it would become established, he reasoned, by dint of being used in the dictionary. "If we consider the English speaking population of the world, it is safe to say that not one person in a million would know what was meant if we should use 'saturation' or 'chroma' or 'purity' (without explicit definition) in place of 'hueivity.' . . . It is therefore futile to speak of preferring an '*established*' terminology in this connection: there is none." This is exactly the wrong tack to take with a lexicographer, whose entire job is to give established terminology explicit definitions. They're liable to get up from their desk after reading that twaddle and make some explicit noises of their own.

assorted sampler of other scientific-ists meant Carhart didn't have time to care.

Now that Godlove was engaged on the color names, Carhart was hoping that with enough prodding Priest would spit out all the other color terms that Merriam was desperately waiting for. Instead, Priest demanded (without explanation) that Merriam promise six full-page color plates for the color section before he began the "color" article that was supposed to have been finished and in Carhart's hands six months earlier.

There is a photo of Carhart on the staff page for the *Second*. His large, sharp eyes are framed by thick wire-framed spectacles; the knot of his polka-dot tie sticks straight out from his collar as if snugged a hair too tight. He looks sensible and smart, not the sort to rave and throw tantrums at the smallest inconvenience. I share this so you will have context for Carhart's response to this ultimatum. Instead of noting the point with the small, tidy check mark that he traditionally drew in the margin next to each paragraph, equidistant between the edge of the paper and the start of the text block, there is just a long dash running from a small blot near the text block itself and trailing off the edge of the paper, a clean representation of the path his patience with Priest had taken on its way out the door, never to return. Merriam had initially contacted Priest in 1929, and in the space of a month he had agreed to define the colorimetry terms for them (about seven hundred words, all told) and write the article on color. Then he wanted charge of the color names; then not. It was now March 1931, and the production of the rest of the dictionary was in full swing. Almost all the other consultants had finished their work; the only areas that hadn't seen any substantive motion forward were color names and colorimetry. The in-house staff, which numbered three dozen, had tens of thousands of consultant-written definitions to get into shape, in addition to tens of thousands of general vocabulary entries to revise and write, and all of it hopefully in the next eighteen months, give or take. And now not only was Priest telling Carhart that he hadn't really *started* the work yet, but he was holding it hostage for six full-color plates.

Color plates are the full-page, full-color inserts of pictures or illustrations that are a luxury feature of unabridged dictionaries (and this very book; see, for instance, page 34), and they are not cheap. They had to be printed on heavier stock than the rest of the dictionary, and that with a glossy or semigloss finish, which of course costs more. Most of the plates required a custom-cut onion-paper leaf over the color print so the inks didn't fade or crock off on the facing page, rendering that carefully composed, crisp six-point type into a smeary mess. In the *Second,* color plates were printed on only one side to ensure that only one flyleaf was required per page, not two, to save money and keep the inks from bleeding through the paper onto the other side. (Good call, because when I open my copy of the *Second,* it's clear that they did.) Because the process of setting and printing them was so costly, most subjects of color plates—birds, gems, flags, things that would benefit from color images and were of arguably greater commercial value than a chunk of colors—got only one color plate if they were lucky. They had already granted Priest two plates, which was, in Merriam's eyes, a princely gift.

There was an additional concern, especially for the color nerds: Though color printing wasn't new, it also wasn't terribly precise. Different ink runs and paper stock, new processes, illustrations of varying chromaticity, printings that were years apart and so faded at different rates, all conspired to make it nearly impossible to produce thousands of color plates of uniform hue, chroma, and value over the lifespan of a dictionary like the *Second*. Priest had already made noise about how he had commissioned (on Merriam's behalf, paying more than they wanted) the painting of a solar spectrum from one particular artist because of the unwavering accuracy of that artist's color work. It was going to be difficult enough to render that painting accurately on the color plates Priest had been promised; they'd probably have to find a special printer who printed to scientific tolerances just for the damned spectrum painting. The thought of producing *six* plates, each subject to Priest's rigorous (and snail's-pace) standards, was the breaking point. Carhart gave up on Priest ever turning anything in and sent a letter to Godlove asking him to, in addition to defining the color names, write the article on color.

Unlike Priest, Godlove dove into the work with apparent cheerfulness and drive. The questions he sent along to Carhart were of a completely different tenor from Priest's: detailed, stemming from the work at hand, and frequent. Should he consider at all which colors were entered into a competitor's dictionary, since it was the most up-to-date and comprehensive reference vis-à-vis color so far, or would that look bad since the dictionary was in direct competition with Merriam? (The latter, and that he knew enough to ask shows more sense than some lexicographers have.) Each page of the *Second* was going to be split, with the main entries taking up most of the page, and then rare, obsolete, or derivative terms being printed at the bottom of the page, under a thick rule and in tiny type. He was planning to put the rare and obsolete color names that have historical significance in that bottom-page chunk; is that a good idea? (It is.) In reviewing the old entries Merriam had sent along to him, he found inconsistencies in how things had been edited and wanted answers before he moved forward: Why was the definition of the color "admiral" "= logwood" when the definition for the color "Algerian" is "the color tanbark"? These say the same things, Godlove reasoned, so shouldn't they be styled the same way? (They should be, and he was doing better than the Merriam proofreaders, who had been the ones who created the confusion while doing their editing pass.) He noticed that Merriam edited one of the draft definitions he sent in for "amber white," but shouldn't "yellower than beige and putty" be reverted to "yellower than beige *or* putty," since the "and" implied that beige and putty are the same color? (He has now worked his way up to overseeing and copyediting the work of the proofreaders and editorial assistants, congratulations.) Godlove even joked like a lexicographer: In noting that he had already done a good deal of the color names defining, he finished, "In doing so, I have felt that I have accomplished practically the impossible. (I have become quite used to doing the nearly impossible in recent months.)"[8]

Godlove's sensibilities were so dead-on that Carhart treated him just like an in-house lexicographer—which is to say, he wrote often reminding Godlove of the deadline, sent him short notes asking him to clarify earlier definitions, and otherwise ignored him. This wasn't because Carhart was overworked; rather, it was because lexicography happened in quiet solitude at Merriam. Carhart had answered his questions more than adequately, he felt, but after much pestering, Carhart finally answered in full—or, rather, "in full," since he still managed to avoid answering many of Godlove's more in-depth questions. The only things he was definitive about were that Godlove shouldn't copy (even inadvertently) any definitions from another source, and that Godlove should give as much of the work to his assistant as possible, since this was going to cost the company quite a bit of money. Carhart apologized for the long response, which was still roughly half the length of most of Godlove's letters to the office.

Buried there in the piles of correspondence from Godlove was the breadcrumb trail that eventually led to a revolutionary change in defining color. If his stint organizing a color exhibit at the museum had taught him anything, it was that scientists and specialists had a very particular way of thinking about color—one that read to regular color-viewing folks as experimental at best or iconoclastic at worst. Point to a tangerine and ask Jack Q. Colorviewer what color it was; Jack would say "orange." Ask a specialist what color it was, and you'd get a whole disquisition on color from that particular specialist's view. Biologists would start burbling on about rods and cones; neurologists would go on about psychophysical sensations and stimuli; physicists would talk about reflection and refraction; Albert Munsell would steal your tangerine, toss it under a colorimeter, and tell you the tangerine was 2.5YR 6/12 while handing you a now thoroughly manhandled tangerine. None of these answers are wrong; it's just that none of these answers are helpful.

Godlove understood that if you wanted to begin educating people about color, you have to start where they are. He began by immediately dumping Priest's "hueivity" on the grounds that it was utterly unknown (and slightly ridiculous). Instead, he went back to Munsell's "value" and "chroma" but, after some cogitation, decided that while "chroma" was absolutely correct, most people used "saturation" instead, and "saturation" would make it clear that they weren't cribbing from Munsell. He chafed at Priest's demand that the core colors used in definitions be only red, yellow, green, and blue, and the resulting Munsell-ish combinations that would have to be used to describe all the colors that fell between red, yellow, green, and blue. "Orange" would be much simpler for the layperson to understand, even if it lacks the scientific rigor of "yellow-red." Nonetheless, he will accommodate the elder statesman. To a point: We don't use compound color terms like "yellow-red" in everyday speech, but adding an "-ish" to the first color? Completely normal sounding—and "yellowish red" preserved the scientific notation that Priest demanded.

The rest of the defining formula also gave Godlove the yips. The formula that Priest had dictated was a concession on his part: The indefinite—nay, *sloppy*—way that color names had been used by nonspecialists necessitated a certain amount of (much detested) wiggle room when defining a color for general audiences. The exact spectrographic composition of "amaranth pink" would, after all, be lost on most users of this dictionary. But Godlove understood that a color with an unrecognizable name couldn't be defined by references to saturation and brilliance—two technical terms—and communicate anything clear to the average dictionary consultant. People wanted a tangible reference point: Saying that the color "émail"* was a blue green of low saturation and medium brilliance was only so helpful. Was it like turquoise, which tended to be bluer, or sea green, which tended to be greener? Was it darker than both of those? Lighter? Washed out or intense? In the absence of actual color chips (the subject of which was not broached for fear that Priest's ears would itch and he'd come crashing back into the project to demand even more color plates), the more information you gave a user in plain language that they understood, the better. Godlove thought it would be better to give a dictionary user a common reference color so they can know just where in the spectrum émail sits. He proposed that they follow Priest's formula, but that they also add a second sentence to the definition: "a [color category] that is of higher/lower saturation and higher/lower brilliance than [a commonly known color]."[9] If "émail" or "amaranth

* Yes, really: The color name is borrowed from the French word for "enamel."

pink" could be defined in relation to other, more common colors, like teal or rose, wouldn't that be more helpful than saying émail is a bluish green, or that amaranth pink is a bluish red? This may seem like common sense to us, but it was a radical approach that neither the generalist would have been able to come up with nor most scientists would have agreed to.

That's not to say that Godlove didn't have his moments, and that Carhart didn't always sit on the edge of dyspeptic agita whenever he received another hefty package of correspondence from Godlove (especially when they showed up sans definitions). For instance, Godlove had decided that the line between color and pigment was thin enough that he could—and should—just traipse right across it. "Ultramarine" was a pigment—a colorant made from ground lapis lazuli—but the word was also used to describe the color of ultramarine. Should the color definition really be separate from the pigment definition, then? What chromatic horrors might arise if we let the chemist drafting the ceramics definition of, say, "zinc pink" (a glaze made from feldspar, silica, zinc oxide, and kaolin) mention that it fires to "a pink" in an oxidization firing, but turns "green" in a reduction firing? What pink? How green? How dare! For Godlove, it made sense to have the color specialist handle both. But Carhart already had another specialist who was handling the pigment definitions—why would Godlove think Carhart wanted him to take on extra work, thereby costing the company more time and extra moolah to boot? It was exactly the sort of overdone undertaking that led Carhart to run hot and cold in his correspondence with Godlove.

But when it came to editorial matters—matters of *real* business, in Carhart's view—he picked up the pace. The new way of defining by comparison would mean that Godlove was going to define by color category and not alphabetically; it would go faster overall, because he'd be using the same comparison colors for each related color, so no need to reference a long list of comparison colors by hue when you went from "amaranth pink" to the next alphabetical color, "amber." Sensible—but it was going to royally fubar the letter-based editorial schedule. "Amaranth pink" was not going to be defined by the time that the letter *A* was set to be proofread and sent off to the printer: It was, after all, a bluish red, one of the last categories that Godlove would define. But Godlove was sanguine: In a letter to Carhart, he noted that he and his assistant had already figured out how to define most of the two-thousand-plus colors that were to be entered into the *Second* in reference to about seventy or eighty "core" colors. The only concern, Godlove noted several times, was that this approach might take more time, but "a great deal of such work has already been done."[10] That last clause was music to Carhart's ears. He didn't ask for clarification, and Godlove, assuming Carhart understood that definitions wouldn't come into the office by letter, didn't offer any.

The pace at which Godlove worked, considering he was also in the middle of his fourteen-hour days at the museum, was well past blistering and moving well into three-alarm-fire territory. He sent lengthy letters almost every other day, asking questions that were detailed yet wide-ranging in their implications for how all the sciences were going to be handled in the Big Book. If color bled into every part of our lives, so, too, did the science of color. Meanwhile, the defining slips came in to the office in overwhelming yet infrequent chunks: All the color names he had already defined (with an emphasis on those in *A–D*), along with dozens of slips for terms well outside Godlove's purview—physics, optics, art history, chemistry, biology, astronomy—showed up in April, giving Carhart and his subeditors about a hundred slips to sort through. Before they could even begin, another packet of letters flurried into the office, asking more questions about areas that weren't color, with more defining slips. Godlove assured Carhart that he hoped to "swing into normal speed" any day now, news that should have reassured Carhart that they'd make up for lost time, but which likely only succeeded in stressing him out even more. It was true that the color work lagged behind the production schedule, but so had most of the rest of the outside consultants' defining work. Carhart was eyeball-deep in reading, editing, and cleaning up the backlog of definitions he should have put to bed months ago. He hadn't even had a chance to sift through all these late-coming definitions, let alone read or think about them. This was the problem with outside consultants: They had other work to do, which meant that the dictionary writing was often shunted off to the side or held for weekends or evenings when one could have a leisurely scotch and perpend the essential nature of the object of one's entire life of study. Or, at least, perpend until one got sleepy from the leisurely scotch and needed to take a thoughtful, mulling nap.

But Godlove wrote as if he had nothing else to do, and thirty-six hours a day to fill. The defining slips still came in, about once a month (too infrequent); still he sent lengthy letters every other week, in addition to near-weekly invoices for himself and his assistant (too frequent). All of the responses that Carhart sent along with the payments included some variation on the theme "We will respond in full when we have more time," but it didn't stop Godlove. May 9, 1931: A letter with seven numbered points, most of which are just FYIs, but tucked in one of the two lengthy paragraphs explaining exactly his methods for measuring and defining the colors is a note that he's found so many errors in the reference he's using that he just might have to start again; how does Merriam want him to handle these errors? Three days later, May 12, 1931: Another letter, this one with fifteen points that need response, including "the question of blue-green or green-blue, discussed on April 22nd" (question 4a) and what he should do if he hasn't received notification from the office that his definitions have been accepted (question 4). "I do not wish to deluge you with correspondence," he writes, then proceeds to send more letters on May 17, May 19, and May 21.

One of the things you learn quickly as a lexicographer is that when the project deadline begins its slow, jerky climb to the top of the roller coaster that is dictionary production, the true nature of the lexicographer emerges: white-knuckled, staring straight ahead, taking one deep breath before the plunge into schedule madness. Carhart was a seasoned pro; he could see, from his juddering seat, that he had only one last moment to try to rein all this back in before he was plunged yell-first into the chaos of trying to see this dictionary to the end of its run. He was concerned; in going through the reams of correspondence and the defining slips, he began, with horror, to see that the forest being created from all these individual trees was not the kind of forest that he thought it would be. It was time for some judicious pruning.

A Brown Study

Good lexicographers are, generally speaking, not people people. You spend your days alone at your desk with your pile of index cards, and when you have dealings with people, it's generally the boss telling you to work faster, or your neighbors asking why you look so pale, or your banker asking why you're constantly overdrawn on your accounts. English can be a right bastard sometimes, but it doesn't talk back to you or require that you do anything but stare quietly.

Carhart was an *excellent* lexicographer.

His first actual response to Godlove, sent May 26, 1931, is absent all the earlier soft soap he had used on Priest and, early on, Godlove. He was, after all, under a tremendous amount of stress: Godlove wasn't his only charge, and he had in-house editing duties to do as well. If he's all dreary, bland business, it's just because that's what life has been reduced to. Carhart ticks off, efficiently and without flourish, answers to the dozens of questions that have come in over the course of the last month, but not without a little expectation setting. Godlove may be the expert, but Carhart was the managing editor here. Future questions about colors should come on the slips with their definitions, not in letters; Godlove should stop messing around with non-color definitions; if they didn't get back to him right away, well, they were busy. "We have many other subjects to deal with in the course of a day," Carhart writes somewhat wearily.[1]

Godlove did not receive the criticism well. He was particularly peeved at Carhart's insistence that he work only on color terms. The real problem, he insisted, is that everyone apart from "an astoundingly small number of persons" believe that color is a discrete field:

> I can say this in all good taste, since only a few years ago, as a chemist and physicist, I was in exactly this state of ignorance myself. I am constantly reading and hearing statements by physicists and other scientists which deal with the subject of color which can only be said to be patently untrue. Such remarks come even from Nobel prize winners, and others of equal repute.[2]

It is incredibly easy to read this statement and marvel at the hubris and presumption of flat-out saying that editors of *The Handsomest, The Best, Webster's Dictionary* are "wallowing in a slough of confusion," as Godlove put it, when it comes to color—a thing that every editor on staff has some familiarity with, if only by dint of being able to see (or not being able to see) it. It also shows just how much Godlove's approach to color deviated from the established mode of Priest, who not only stayed in his (very narrow) lane but resisted all attempts to be swerved out of it. Godlove didn't just care about color names and colorimetry; physics, chemistry, optics, art, manufacturing, and even cooking all touched on color. Therefore, the color man should touch on all these topics as well.

If there was any one example of Godlove's sense of overreach, it was the entry he had turned in for the word "color." It was not a normal dictionary entry, but a multipage article that gave a complete survey of the history of colorimetry and laid out Godlove's—and now Merriam's—philosophy of color. What color was, what color was not; how color was measured, and why measure it at all; what light was, what vision was; how chemistry, physics, and optics jostle up against each other in colorimetry. It is a single-spaced academic paper with references, and it was far, far too much for Carhart to even read, let alone edit. He enlisted his colleague Mabel Martin, who knew something about color, to help.

Dr. Mabel Martin was a rare bird among dictionary editors: a (1) woman scientist who (2) had held down actual jobs in industry before becoming a lexicographer. Her credentials rivaled those of the higher-ups, like Carhart, on the Merriam Editorial Board: bachelor's from Mount Holyoke just down the road, master's and doctorate in psychology from Cornell, elected to Sigma Xi (an international honor society for scientists) upon the completion of her doctorate in 1922—one of the four women out of thirty-seven new inductees at Cornell. She went into academia and consulted with various psychology clinics until 1929, when she landed at Merriam in time to work on the *Second*.* Her dissertation, "Film, Surface, and Bulky Colors and Their Intermediates," teed her up to be the in-house expert to edit the out-of-bounds expert.†

* But not for long. She left the syntax mines after five years of full-time drudgery and went back to teaching and clinical practice. But her reference career didn't end in 1934; she wrote the psychology-related articles for Funk & Wagnalls' *New International Yearbook* from 1936 through the mid-1940s.

† Martin is a ridiculously common name, but Godlove knew from the comments on the article that someone with knowledge of color had revised the entry, and he asked if *this* Martin was *that* Martin, the Dr. M. Martin who published the paper "Film, Surface, and Bulky Colors and Their Intermediates" in *The American Journal of Psychology* in 1922, and "whose paper . . . I read with interest several years ago and reread with great profit recently." Game recognize game.

The bulk of her changes turned Godlove's scientific dissertation into plain English and focused the article on the kinds of information that would be helpful to regular people and not scientists. She removed his (excruciatingly abstruse) definitions for "light" and "visibility" because even scientists argued over their technical meaning, and the average dictionary reader didn't need to be further befuddled. A section about secondary colors and "mixing colors" was added; the idea that red, yellow, and blue were *the* primary colors that mixed together to make black was scientifically, provably wrong,* but it's the sort of information that people would expect to see in an article about color. Martin revised out a long note that material objects didn't *have* color but were merely perceived by a viewer to have a color, because it made as much sense to the average reader as airborne fish, which is to say, not at all. Nearly every paragraph of Godlove's essay was revised to be accessible and readable by the general public—not the color scientist.

This review was not typical; it was the result of some in-house brouhaha. The turd stirring over the "color" article started when Asa Baker, the president of Merriam and a member of the editorial board, started nosing around the special-correspondence files to see why production was behind schedule. He found Godlove's "color" article and flipped through enough of the correspondence to get a sense of Godlove's wide-ranging scientism. It was, evidently, the focal point of all his ire: Baker decided that therein was everything wrong with the entire approach to special editors that Merriam-Webster generally, and Carhart specifically, had taken.

Baker was, as the editor in chief William A. Neilson would later describe him, "a dictionary man" from "a dictionary environment" with "dictionary sense."[3] He had done editorial work at Merriam-Webster before being elevated to the presidency of the company,

* Then what does science say? If you don't know already, you'll have to read (or skip to) the epilogue to find out.

so he had fixed ideas of what exactly constituted a dictionary. Baker, it should also be said, was responsible for the finances of the company. The *Second* would eventually cost the company $1.3 million to produce. That's just shy of $25 million in today's coin and, even in the dictionary-hungry days of the 1930s, substantially more revenue than most dictionary publishers would be able to produce in a dozen years. He had to account for every penny spent to the board of directors. It made him a notoriously brusque man.

The initial "color" article was bad enough, but it was one aside in Godlove's (admittedly) conciliatory response to Martin's edits where he offhandedly supposed that for a general publication common usage had to prevail over good science that really did Baker in. After reading both, he fired off an interdepartmental memo where his deep irk would be memorialized. Good science and good general usage were not, in fact, at odds with each other. It had been his strong opinion decades earlier that the unabridged be supersized into a dictionaripedia to satisfy public demand, and that dictionaripedia should present clear, concise facts about the world and the language provided by the best scientists dictionary money could hire. But those facts, as Baker saw them, should be sensible: How could some puffed-up scientist insist, for instance, that facts tell us that objects don't have color when the entire written record and Mr. Asa Baker's own two eyes contradict him? "We have won our success by relying on specialists, and have secured a remarkable list of modern scholars," he groused, "but these men will not co-operate with us if we attempt to overrule their convictions as to <u>facts</u>."[4]

Baker's typewritten temper tantrum wasn't really about Godlove at all: It was about the impossibility of lexicography in an age of progress. In 1929 alone, the year that Carhart first approached Priest, Edwin Hubble discovered that the universe kept expanding, the cell coenzyme adenosine triphosphate was discovered and estrogen was isolated, William Frederick Gericke demonstrated that plants didn't need soil to grow to maturity, color television was first publicly demonstrated at Bell Labs in New York, coaxial cable and an early fax machine (called the Hellschreiber) were patented, human electroencephalography was discovered, the first ring IUD was invented, Alexander Fleming published his first paper on the weird antibacterial mold he accidentally discovered (called penicillin), cheap celluloid sunglasses were manufactured for the first time and sold in Atlantic City, and Polaroid polarizing film was patented. That was just one year, and this small list of discoveries affected thousands of entries in the *Second*. When so many fields of study are growing by leaps and bounds over the half a dozen years in which you are writing your authoritative doorstop, how can you possibly keep up? At a certain point, the publisher just had to close the books, financially and intellectually. And when it comes to something like color, which their own specialist admits is still very much in flux, is it worth all this wasted time and money to print an extensive article that may, by the time this boondoggle on India paper gets put out in the world, be not just out of date but *wrong*?

And who's to say that Godlove was even approaching "right"? Are you seriously going to tell people that the most foundational thing they know about color—that when they look at the expensive dictionary, they can see that it is brown—is wrong because their eyes are literally deceiving them? That material things *aren't* a color; that science believes that light as it is reflected and refracted off/through/in/around surfaces is the thing that causes a psychophysical reaction between the eye and the brain that we apprehend as a sensation called color? The fact is that Baker's shouty face *was red,* and not that Baker's shouty face was colorless but had absorbed all wavelengths apart from the ruddy ones, which exist only in the neural network of a shrinking Carhart. Baker's terse one-page memo offers three options for moving forward: print this monstrosity of an essay unedited (unacceptable); find another consultant who would be more pliable (unthinkable); or "show Dr. Godlove that he is wrong, and we are right" (unquestionable).[5]

Had Godlove seen this interoffice memo, he likely would have sworn a blue streak—especially since he had agreed to all of Martin's edits, and pretty damned cheerfully to boot. But the memo was never intended for Godlove; its effects on its target, Carhart, were immense. Baker was one of the chief architects of the dictionaripedia; he knew full well that the only way to produce a dictionary this large and this scientifically rigorous with an editorial staff made up of a couple dozen English-lit nerds was to hire talent in; he had been an editor when this scheme started and understood that the role of the in-house lexicographer was merely to collect definitions from the consultants and hammer them into a dictionary-shaped object. To change the approach to the specialists, at this late stage in the game, was madness; to do it in such a scolding and bullying way was degrading. Carhart's response to this memo, and the evident dressing-down he got for not keeping a tighter rein on these madmen called special editors, was one of wounded confusion. Since the memo had been written, and so entered into the "record" of the *Second*'s production, Carhart made free to comment on it (though not in a memo sent back to Baker; he merely annotated the memo and filed it away for posterity). Why had Baker gone over his head to Thomas Knott, the general editor, to rework Godlove's article when Carhart had been the one who had already laid the groundwork? Why was Baker insistent on naysaying the specialist? Godlove seemed to be happy with the huge changes made to his article, which was a point won, in Carhart's view: why the foot stamping on Baker's part? "There seems to be a serious difference of opinion," Carhart writes, but comes no closer to why the difference exists, or what to do about it.[6]

Baker, the keeper of the purse, had spoken, however, and made it clear this was no time to enter into metaphysical discussions on the nature of "light" and nurse wounded feelings (which, Baker would surely point out, were the result of Carhart's own missteps editorially). We have a dictionary to produce. Quit all this farting around and produce some page proofs. Carhart, bloodied and bowed, complied.

For a while, Carhart's upset infected his responses to Godlove. "We find our time is exhausted for this afternoon," he says in one reply to another sheaf of correspondence, promising to answer the rest of Godlove's questions the next day, which he does not. Maybe it was the production schedule, or maybe it was that the stack of correspondence from Godlove was a constant reminder that the president of the company has, at this late stage, decided that perhaps hiring all these specialists, which had been Carhart's occupational raison d'être for years and the reason for his promotion, was the worst idea possible. Carhart cannot muster more than a few sentences in response to Godlove for the next few months.

Then the impossible happened: Godlove's correspondence dropped off so precipitously that Merriam worried he was going to pull a Priest and flounce. Carhart didn't begin to panic until the days without a Godlovian tome turned to months. Nothing could entice Godlove to respond: Have you thought about the color plates/the color article/the rest of the definitions? Radio silence. Be careful what you wish for.

On November 17, Godlove finally responded to Carhart's increasingly nervous check-ins. "I am sorry that I have been unable to make the color work progress more rapidly recently," he apologizes. His wife, he reveals, is dying. "As she is suffering a good deal and is loosing [*sic*] strength so rapidly that she can not continue much longer, it has been very difficult for me to concentrate on my work. I have, nevertheless, put in quite a lot of time at work, though I have not accomplished much," he explains. The remaining two pages of his response are business as usual, though they are a good deal less theoretical than his correspondence had been. He's been shifting around color chips, noticing color names in magazines and newspapers, counting again and again the number of times that his core colors have been used in color definitions (about seven thousand times, by his count). He assures them he will not bill them for this, because this is "the most mechanical kind of routine hack work." "I am very sorry to have held up progress," he finishes, "but I have worked recently under very trying conditions, and hope you will forgive me."[7]

This letter is jarring. He was, after all, a specialist: work for hire. No one in the company knew anything about him—sometimes even including where to reach him—nor did they ever ask him anything about his personal life. They had no idea he was married; to have asked would have been improper, far outside the bounds of this lexicographer/consultant relationship. Yet Godlove's candor and numbness are on full display in his letter. He's writing as if Carhart were a longstanding colleague, a brother-in-arms, someone he might have grabbed lunch with and seen around town—and not someone who had hired him as a freelancer and then had let him twist in the wind with only the most perfunctory and formal responses to all his letters. It's entirely professional and tenderly human.

I.H.'s wife, Esther Hurlbut Godlove—whom he had met when he was a professor in Oklahoma; whom he had married in 1923; who moved with him to Baltimore when he took a job at Munsell in 1926; who gave birth to their son, Terry, in 1927; who was an accomplished musician and participated with I.H. in the church choir; who played an embodiment of spiritual Grace in an Easter pageant just months earlier, when I.H. was busy puzzling out "aloma" for Merriam-Webster—died November 21, 1931, of breast cancer.

Carhart did the needful and responded to Godlove's letter with a note of condolence, but then was right back to business: When can we expect the rest of the color definitions, have you attended to the notes we've sent along, we really need to get the color plates taken care of, you know, where *are* you, our letters to New York were returned? (Godlove was, in fact, staying with his in-laws in Washington, D.C.) It was not just lexicographical awkwardness—or, maybe, not entirely lexicographical awkwardness. Carhart, already shriveling under Baker's incandescent glower, now had in Godlove's letter a sharp and present reminder of the sudden death of his first wife thirty years earlier. They had been married only three years; she died in Germany, where he was studying phonology—at Merriam's behest. He reveals none of this to Godlove, a brother-in-grief. He apparently reveals none of this to anyone.

Godlove's responses are dull and perfunctory—mostly invoices except for one letter, which comes in two parts and after a significant amount of typewritten hand-wringing from Carhart. The first part of his letter answers some perfunctory questions about the color plates; the second part is a copy of a letter he sent to Charles Bittinger, the artist who had been contracted by Priest to paint part of the color plates. He sent this copy as surety that he had communicated with Bittinger in plenty of time. He possibly also sent it to remind Merriam and Carhart of what sort of disruption he had suffered:

> When I finally started answering your last letter, I saw with a terrified shock that it was dated March 4th. Although I have kept busy, it appears that the blow which fell on me last fall, has caused me, as one result, to lose all consciousness of time; and I am going around in circles, like a chicken which has just lost its

head. Time does not seem to make matters better, in fact just the reverse.[8]

One of the pitfalls of any highly detail-oriented work, like lexicography, is that you are sometimes so mired in the picayune scrambling of putting a dictionary together that you forget that life exists outside that eleven-by-seventeen-inch galley page. Godlove, living in Baltimore, had been handed the exhibit job in New York, which took him to the city for a year (and, we can guess from the hints left in his correspondence with Merriam, away from his wife and toddler son). When that was done, his wife was almost immediately diagnosed with breast cancer, which necessitated his swift relocation to the mid-Atlantic, where Esther's family was and where she could get treatment in Washington, D.C. She died only a few months after her diagnosis, leaving him a single father of a four-year-old son; a man with no steady employment (particularly as the color defining at Merriam-Webster was being rushed to its inevitable conclusion); a widower who was unmoored and undone by his wife's sudden death. If he didn't have the emotional wherewithal to continue to respond to Merriam immediately, giving his very marrow for the sake of Mother English and all in a tone of collegial bonhomie, no person of feeling could blame him.

Unfortunately, he was working with lexicographers, who were unsure of how to approach those pesky "feelings."

Carhart and Godlove were both sinking and took solace in the work that meant the most to them. Carhart reverted to his baseline chiding; Godlove essentially abandoned Merriam. When he did respond, he met Carhart's apparent irritation with gloom—an emotional state that is practically the lexicographer's stock-in-trade. "I regret exceedingly," Godlove writes, "that the still unsettled condition of my feelings has kept me on the move and has made correspondence difficult, with the result of considerable inconvenience and annoyance to you." He then indulged (uncharacteristically) in a little snark: "I assure you that had I anticipated any such eventuality in the spring of last year I should never have accepted the engagement."[9]

The defining finished in 1932, and the plan for the color plates in 1933, but the notes from Carhart kept coming: We received a letter from a person we've never heard of claiming that we should enter the word "varial" in regard to color; please answer him. We are going through our list of entries from the 1925 copyright of the *New International* that we cut up and pasted onto index cards for you, and find we are missing five, and though we sent you thousands of such slips, we require that you find and return these five immediately. Please send back the copy of Maerz and Paul that we lent you, even though we told you two years ago that you could actually cut it to pieces and use it for mocking up the color charts. Godlove had wound down, but Merriam was jumping: proofreading, bluelining, more proofreading, and in general wrapping up the great and difficult work of the *Second*. The end of the roller-coaster ride was in sight, and no one felt the rapid approach of the end faster than the managing editor for the whole paper-jammed megillah, Carhart.

Many people conceive of the creation of a dictionary as the best sort of intellectual exercise. The editorial floor is like a busy beehive, drones flying in and out with pieces of language stuck to them, workers busily converting them into slips and definitions and more slips, and tucking them away in their cubbies. And at the center, the queen: Mother English, producing more work and more workers, keeping the hive humming. This isn't a bad conception of what it's like to create a dictionary, but it's an anodyne one.

Creating an unabridged dictionary is, at its core, a lesson in futility. Everything works against you. English, as a living language, is always on the move, and lexicographers are doomed to jounce after it, winded, their entire lives. They can't assume that they know where language is going: It jinks and prances off in unpredictable directions. This means that the minute you decide to start carefully delineating how a word is used—the minute you squintingly put pencil to defining slip and begin scratching out a definition—becomes the point in time when you must ignore how that word is currently changing and how it will change (between now and the next edition, at any rate). The stated claim of an unabridged dictionary—that it encompasses all of the language and is, as one marketing blurb put it, "the sum of all human knowledge"—is a lie that the lexicographer has to live with all their days.

Then there is the sticky reality of the hive itself and the workers therein. Samuel Johnson defined "lexicographer" as "a harmless drudge," and while most people take that definition as evidence of Johnson's self-deprecatory wit, it is as serious as the proverbial heart attack. The process of writing a dictionary is filled with enough interminable slogs through redundant processes and abstruse minutiae that it makes the most staid, pencil-pushing government bureau look like Coachella on ecstasy. This is an industry that welcomes its shiny-eyed new employees with a forty-seven-page memo on punctuation that is so convoluted and esoteric it would make God weep lightning-hot tears of boredom—and continues in that vein until these tender young scholars are broken down into a pulpy mass of editorial promise, where they are lovingly masticated by their ruminant industry for decades. Dictionary editors do not get to claim ownership of the work of their hands; dictionaries are authored under a company rubric because no one editor can (or should) make or break the production of a dictionary. Just as English is a democratic institution, compiled by the nameless masses, so are the reference books that document it.

When the work of writing dictionaries becomes daily drudgery—when the low pay and solitary nature of the work sit heavily in your stomach and begin to gurgle a hole through it—management will remind the downtrodden lexicographer that it is an honor to do this work; that very few people get to experience the joy of tracing the meanderings of English. And that is true: Those moments of serendipitous wonder sparkle like crystal. But the longer you do the work, the less likely it is that you get to keep doing the work. Dictionary publishers need editors who can manage people and schedules and budgets; editors who can train younger lexicographers how to parse the fine-grained wonders of "as," then go straight into a board meeting and argue to preserve the budget that keeps those younger lexicographers employed and pondering adverbs; editors who can call (Call! On the phone!) compositors and binderies and explain to them the dire history of the word "deadline" and its applicability here.* Senior editors are the ones given these tasks, because management does not trust those tender slabs of learned inertia staring flabby-mouthed into the middle distance waiting for the right definition of "bowl" to coalesce; this is work for a seasoned professional. And all the while, as the senior editor attends to the molehills of book production, they can see English just in their peripheral vision, gamboling off in a direction they desperately, quietly long to go. It is a tension that most senior lexicographers manage to hold, usually in their shoulders and necks. But it is a tension that can also break you.

* "Deadline" first appeared in English in the 1860s and referred to a literal line or boundary around a prison. Should a prisoner cross the line, they'd be shot dead. No pressure, compositors/binders.

Noah Webster spoke about his extreme depression and anxiety; Samuel Johnson suffered bouts of the blues. It was common for there to be a glut of resignations after the completion of a big book like the *Second,* which demands everything of you and gives you, in return, a lot of headaches and not enough salary to ameliorate them. Queen bees, it is important to note, can and do sting.

Paul Carhart would not see the publication of *Webster's Second New International Dictionary* in 1934. The end came faster than anyone anticipated; he died by suicide on October 27, 1933.

Carhart's sudden death was reported nationally, but there's very little in the corporate records about it. No memo to the staff extolling the hard work that Carhart had done to get the *Second* off to the printer, no carbon copy of the condolence note sent to the second Mrs. Carhart, no expense receipt for flowers sent to the funeral or the family. It might have been superstition; it might have been busyness; it might have been guilt. Mrs. Carhart reported to police (who, in turn, reported it to the papers) that she was not aware of any motive for his suicide. He had always been a nervous man, she noted—and with a boss like Asa Baker, who wouldn't be?—but he "had not appeared morbid or abnormal," as one newspaper report put it.[10] The *Times Union* of Brooklyn, New York, heard something different in Mrs. Carhart's report that her husband had been nervous. "Paul W. Carhart Reported Victim of Overwork," reads the subhead. "Carhart was said to have been extremely nervous from overwork," the incredibly short and gruesome obituary ends. "He leaves two children."

Godlove's condolence letter over Carhart's death is heartfelt. "Though my contacts with Mr. Carhart were limited to business," he writes, "I regarded him as a man of great ability and character, and held him in affectionate regard, so that I am able to feel a measure of the loss and sorrow of his associates."[11] He extended his sympathy to the other editors in the office, and then he promptly disappeared. The last correspondence in the Merriam files with Godlove was from 1934, the year that the *Second* was published. Merriam had asked if he would send them a photograph they could use in the promotional materials for the dictionary. Since the *Second* was to have a number of color plates, they figured that they might as well include photographs of all the consultants as well. Godlove sent no reply; his picture is nowhere to be found in the promotional materials for the *Second*. Whatever joy he had taken in lexicography, he had other things to work on now. He had grown tired of their haranguing.

But old habits are hard to break. Eleven years later, Godlove wrote back.

The Silver Bullet

Life after the *Second* had been good to Godlove. After Esther's death, he established his own color consulting business in Washington, D.C., to be near his son, Terry, and Esther's family, but it was unsuccessful ("as expected," writes I.H. in 1946 to Merriam: There was a depression on). But in 1935, he landed a job at the DuPont Organic Chemicals Division's Technical Laboratory in Deepwater, New Jersey, working as a chemist and physicist specializing in the measurement of color. It was a plum assignment. DuPont was insulated from the financial plummet of the Great Depression, and at that time it was one of the biggest U.S. chemical corporations and had a strong research program. All the money it dumped into physical chemistry paid off big-time: The Technical Lab at DuPont became a workhorse of the Deepwater facility where synthetic rubber, neoprene, and nylon were created. Before Godlove joined the staff, DuPont had also created Lucite, cellophane, and rayon; nearly all of those could be (and, once they became commercial goods, were) dyed. Godlove's hands were full, then: Much of his work dealt with lightfastness, color matching, and standardization of dyes—all vital to DuPont's continued color domination. The work of the color section at DuPont was so well known that a 1935 *Popular Science* article focused on what it takes to get a job there. It was breathlessly headlined COLOR WIZARDS IDENTIFY 100,000 HUES BY EYE.[1]

It's quite the jump to go from curating a museum exhibit and writing dictionary definitions to heading up one of the world's biggest dye works. Godlove owed it all to drugs. (Not like that.) While he was banging out the definitions for "cracker" and "aloma," the U.S. Pharmacopeia (the standardized list of all registered drugs sold in the United States) was mired in its own definitional funk.

We're used to online versions of the USP, as it's called, that include pictures of each compounded pill or tincture in our bathroom cabinet. But images in the USP are a recent addition: Early editions were all text. The earliest editions, like the one that Godlove worked on, primarily listed drugs that we would think of as ingredients. This was the era of the corner druggist, who would whip up a little sum'in-sum'in for your cough, your rheums, your grippe, your fever. The USP included in-depth descriptions of each listed drug, and occasionally those descriptions included colors. These color descriptions were important: They prevented a bumbling or harried pharmacist from accidentally grabbing the brown bottle of *acidum tartaricum* (whitish crystals, laxative) when they meant to grab the brown bottle of *acidum tannicum* (greenish yellow, antidiarrheal). But each iteration of the committee charged with revising the USP used slightly different descriptions for the color of a drug, a preparation, or a chemical reaction, and these terms were sometimes not straightforward. Reports* from the

* Due diligence: I believe that these reports were on an unedited draft of the USP. I read the edition of the USP under examination, and did not see "blackish white," sadly. I did, however, learn a lot about purgatives, laxatives, and which commonly

1920s claim that the USP described one drug as a "blackish white,"[2] which is as helpful as describing steam as a "dryish wet." It was not the only confusing description: The revising committee kept a list noting which color descriptions might be problematic (including the delightful contradictory "reddish green"). Your pharmacist's interpretation of these colors could be the difference between life and death (or, worse, eternal gastrointestinal distress). The USP Revision Committee decided to devise a system that would accurately describe colors in plain English. After about a decade of floundering, they called in reinforcements.

In 1930, the head of the USP Revision Committee put out a call to anyone interested in the science and problems of color to brainstorm ways of fixing this pharmaceutical mess. Physicists, chemists, art teachers, technical instrument manufacturers, makers of paints, pigments, and dyes, authors of textbooks on color, philatelists, ornithologists, horticulturists, fashion forecasters, and colorimetrists from the NBS attended the meeting. All agreed this problem could only be solved in community with one another. What was needed was an interdisciplinary organization that would focus on coordinating all aspects of color and color names.

It was a much bigger deal than it sounds. Academic and technical societies were siloed, even for interdisciplinary fields of study. But color touched everything, and there was no one clearinghouse for all color science, all color psychology, all industrial applications of color, all artistic applications of color. In 1931, delegates met in New York at the Museum of the Peaceful Arts, which housed Godlove's exhibit on color science, and organized: They called themselves the Inter-Society Color Council, and they decided to tackle the USP problem. Godlove, an early member of the ISCC and

used medical chemicals you absolutely should never touch, taste, inhale, or set down too forcefully lest they explode.

someone who was making cash money working on standardizing color descriptions using plain English, immediately joined the subcommittee on Problem 2, as it was called.*

So much of the problem of Problem 2 was that despite the number of color standards out there, none focused on standardizing color *descriptions*. Why bother? Color standards contained color plates full of color chips. You need to know what dusky rose is? No problem: There's a picture. The colors these standards cataloged were also an issue: They focused on color names given to them by science, art, and/or industry, no matter whether those color names make sense to the average person. This leads to all sorts of denotational confusion: Looking through the various standards, regular folks will find that emerald green is not the color of emeralds, horizon blue is actually green, Java is brown but is a different brown from Java brown.[3] And while every industry, every form of art, every branch of science had developed language to describe color, these languages were often mutually unintelligible. The Problem 2 subcommittee, chock-full of scientists, industry workers, and some fashion folks who spoke different color dialects, had to work together and build a framework for plain-English color descriptions from scratch.

Work began in 1932, and Godlove's date-stamped fingerprints are all over the guiding documentation. His six-point explanation of the problem begins not with chemistry, not with physics, not with perception. "*First,*" he ticks off, "in no sense of the word have color names ever been really defined. Standard, unabridged dictionaries are practically devoid of definitions of color names."[4] In other words, don't get your hopes up: There was no recourse to the authorities for language, either. Given that, why not approach

* Problem 1 had to do with adopting a standard for illumination that had been put forward by the International Commission on Illumination, or CIE (an acronym for Commission Internationale de l'Éclairage. Everything sounds fancier in French).

how to describe colors from the science side of things rather than the language side? Science helps cut through the cacophony of different plain-language systems; once you measure a color and chart out its specifications, you can use that to build a language system. He proposed the bridge between scientific notation and just words: A colorimetrist at the NBS would take color measurements of some of the USP's various powders and liquids, and then compare those measurements with Munsell's science-friendly standard. That would help them literally map colors, and once that was set, the ISCC could use what Godlove had learned at Merriam writing color definitions to come up with the ur–defining formula that would take care of inanities like "blackish white" and "reddish green."

Simple in theory, which was where Godlove thrived.

If you're going to come up with a defining system for color, you have to start at the very beginning: What are the most basic terms you'll use to define all the colors of the rainbow? Get ready: Even the simplest color terms can't be taken for granted. Start listing the "most basic" colors you can think of—the colors you'd use to describe other colors—and you'll come up with most of the colors in your kindergartner's crayon pack: red, orange, yellow, green, blue, purple, brown, white, black. Maybe you'd add "pink" in there; ask your friends and neighbors and you might add "gray" in there, too. Congrats: You, your neighbor, and your box of grade-school art supplies have just named the eleven basic color terms in English. In the dictionary biz, we call these genus terms; they are the basic category words that you use to define other colors. "Dandelion" describes a yellow; "lime" is a green; "indigo" is blue.

Scientists and color specialists have been waiting patiently while you tick names off on your fingers, biding their time so they can tell you how very, very wrong you are. First, your most basic color words should be for the colors that are *spectrally* pure—colors that can't be broken down into other constituent colors or aren't a blend of other colors. "Black" and "white" aren't, technically, spectrally pure because they are none of the colors or all the colors, respectively. "Gray" is suspect—it's really just about lightness, not spectral hue. Out all three go. "Orange"? You mean "red yellow." "Brown"? Several color categories at once! Red yellows *and* yellow greens, please be specific. "Purple"? I beg your pardon—purple is not a spectral color, but a trick your brain pulls when your retinas perceive red and blue together. Since it doesn't technically exist on the visible spectrum, it's not a color; it's just a brain wave.*

So we're left with "red," "yellow," "green," "blue," and "pink" as the "okay" basic terms to define with. Not so fast, specialists say. The basic color terms also should be *linguistically* pure. The word used to describe the color should, as its originating point, describe only that color. "Pink," then, is disqualified: It started life as the name of a flower. All pinks are just light reds—or light yellow reds, or light blue reds. Simple enough!

When the ISCC began working on Problem 2, they had assumed they would use basic color terms that most scientists and color specialists generally accepted as basic enough. Munsell was a good model—he limited his main color categories to red, yellow, green, and blue, and threw in purple to accommodate the common understanding of the unwashed masses—and soon the subcommittee realized it was not going to work. If the answer to Problem

* Of course, different color specialists and scientists disagree on the definition of "spectral purity." Some physicists and biologists argue over whether yellow is spectrally pure: You can isolate a wavelength in the visible spectrum that is yellow, but the human retina has only three photoreceptors that are keyed to wavelengths that read as red, blue, and green. This argument is the source of the feud between the nineteenth-century German physicists Ewald Hering (pro-yellow) and Hermann von Helmholtz (anti-yellow), which is worth hunting down because it was cattier than a hot tin roof.

2 was a commonsense rubric for describing all colors using only words, how are you going to convince people who have called all the red yellows "orange" since the sixteenth century that they are wrong?* What good is scientific accuracy if your target audience thinks you sound like a loon and discounts everything you say?

Godlove understood this keenly. At the museum, he had to take highly technical optics, physics, chemistry, and translate that for the average nonscientist *and* make it interesting. At Merriam, he was constantly being pulled between what Priest wanted (scientific purity above all else) and what Carhart wanted (comprehensibility by the lowest common denominator of reader above all else). He didn't see these two desires as diametrically opposed: After all, language is a system bound by rules, and we can describe incredibly complex concepts using this system, as squishy as it is. All you need to do is be a lexicographer about it: gather evidence, sort it, and apply scientific rigor.

So he did. He had been researching the history of color use throughout the ages, so he felt confident that they needed to be flexible in their core color categories. "Brown" was a step too far—despite its long historical use, it was a word that spanned too many color categories—but "orange"? If we want people to trust us, let the people have their oranges. "Purple"? This is not the time or place to get into nonspectral colors and insist on "blue red."

But this is the easy part. In each of those main wedges are thousands of subcolors—how are you going to describe those? Back to common usage. Godlove gathered common color names that

* Indulge me in a little aside: In recent years, it has become popular to say that because we didn't have the word "orange" until the sixteenth century, English speakers couldn't see orange—or couldn't see it very well. This is a load of red-yellow hooey: We actually have a lot of words older than the word "orange" for the color stimulus we interpret as orange. That's like saying that until we created the term "punch line" in the twentieth century, jokes never resolved. Shakespeare would have something to say about that—but nothing funny, since the English were incapable of humor until the word "joke" appeared in the late seventeenth century.

contained some sort of modifier—all the roses, from "dusky" to "dusty" to "dull" to "light" and beyond—took their colorimetric measurements, and began to plot them on the Munsell color space. As he did, he found that there was an identifiable difference between "dark" colors and "deep" colors, "dull" colors and "weak" colors, "light" colors and "bright" colors. Each descriptor occupied a specific zone on a hue plane, like neighborhoods of a city. It was true that English wasn't as precise as scientific measurements, but it also wasn't true that color names in common usage were, as Priest told Carhart, "merely arbitrary labels without logical or scientific basis (and are often fantastic and bizarre)."[5] There were patterns of use that the scientist could adopt to at least get Jack Q. Colorviewer from one end of the city to the right neighborhood.

There were other patterns, too, that signaled to Godlove that some people had a nascent three-dimensional understanding of color, whether they realized it or not. Ask your hypothetical 1930s housewife picking table linens to match her dining room wallpaper: If the flower buds on the wallpaper were pink, she would be able to tell you that a *purplish* pink would bring out the light-blue ribbons on that wallpaper, whereas a *yellowish* pink would clash, but that purplish pink needs to be *light,* because a *deep* purplish pink would overwhelm the lighter colors in the wallpaper. That chunk of light red the hypothetical housewife was navigating contained colors we'd classify as "pink" but which lean in different directions—up and down the value scale to "light" and "dark," along the chroma scale from "strong" to "weak," nodding its head toward one of its hue neighbors (purpl*ish,* yellow*ish*) but staying in its own yard.

It reinforced Godlove's scientific approach to a plain-language color-defining system. He believed—*knew*—that average people had the capacity to understand the nuances of color science. He had watched people visit the Museum of the Peaceful Arts to marvel at the color exhibit, and had seen those same people walk away with some small but new understanding of color science, and be hungry for more. We had grown accustomed to scientific breakthroughs practically every other day—new planets, new stars, new theories of how the universe came to be. We had welcomed scientific innovation into our kitchens, our workplaces, our garages. Science and the quotidian kissed every day. Why couldn't they meet in language, too?

His (incredibly long) memo to the subcommittee laid out a comprehensive rubric for using plain-language descriptions as guideposts for people as they navigated through the color space:

> It is possible to name any color on a scientific basis and with the use of 12 ordinary English words, the adverb *"very,"* and the suffix *"ish."* These words are: *red, orange, yellow, green, blue, purple, light, medium, dark, weak, moderate, strong*. Each word has a very definitive, positive meaning. Whether the color specification be by spectrophotometry, monochromatic analysis, trichromatic analysis, comparison with color charts, use of the Lovibond glasses, or any other standard method, the naming in English words can be scientifically accomplished.[6]

And to prove his point, he included a chart overlay to use with any page of the *Munsell Book of Color:*

light weak	light moderate	light strong
medium weak	medium moderate	medium strong
dark weak	dark moderate	dark strong

Grab a fistful of paint chips, open up your handy *Munsell Book of Color,* and slap this over any hue page, and suddenly you'll be able to describe the colors on the page (and any colors that match it) in systematic plain English. You need to describe to your friend what the "seafoam" paint chip is? Match the chip to the Munsell page and begin decoding. Munsell's 10GY*—the yellow green that tips more toward green than yellow—becomes "yellowish green"; "seafoam" is in the "light weak" square; this paint is a light, weak yellowish green. "Kelly green"? A medium, moderate yellowish green. "Bottle green"? A dark, weak yellowish green. "Lime green"? Light, strong yellowish green. Problem 2: solved.

ISCC members who made it all the way through the memo were impressed with what the rubric's possibilities were. Scientists, fashion forecasters, horticulturists, geologists—what if we have the answer to all our inter-translational problems? Dorothy Nickerson, a colorimetrist at the USDA, wanted to bring it to the broader scientific community. She published a paper on the rubric in 1940 and called it the "ISCC-NBS method."

It wasn't just the museum work that convinced Godlove that the world was ready for his plain-language revolution: He had deepened ties inside the industry and began tracking, with anthropological zeal, color interest outside the industry. In 1943, he took a job in the new Central Research Laboratory of General Aniline & Film, which had only a year before been seized by the U.S. government under the Trading with the Enemy Act from the German

* Munsell's hue notations always throw people. The order of the main colors (red, yellow, green, blue, purple) moves around the Munsell solid clockwise—R, Y, G, B, P in the nomenclature. But the notation for all the in-betweens, like red yellow or yellow green, are notated counterclockwise—RP for red purple, PB for purple blue, BG for blue green, GY for green yellow, and YR for yellow red. There is a very important reason for this, and that is that Munsell thought spirals were really cool.

company I. G. Farben, to undertake research on dyes, intermediates, resins, polymers, and photographic products using seized German patents.[7] He was on the board of trustees for the newly formed Munsell Color Foundation, where he had gotten his start in the color industry; he presented on color at an alphabetic jumble of conferences and published an overwhelming number of papers in industry and academic journals; and he was more and more involved with the ISCC.

This vast sweep of interests and specialties made Godlove an ideal editor for the ISCC newsletter, which he had taken over in 1936, and it was the years of curating the newsletter that really convinced Godlove the world was ready for—nay, needed—*the* Color Standard. The ISCC membership had broadened significantly from its scientific beginnings; by the time Godlove was at General Aniline's CRL, the ISCC membership included color consultants for retail, color psychologists, manufacturers, individuals interested in color theory, artists, dyers, designers, color educators, and technologists from the film and television industry, where color was the Next Big Thing. Godlove wanted to cater to them all. He scoured the news and solicited newsletter items from friends, and the newsletters—like the color fields—boomed. Some members might have found the twenty-page newsletters overwhelming, but Godlove was galvanized by them. In collecting tidbits on color for the ISCC newsletters, he saw a broad interest in what color was, how we saw color, what color did to a person, and what color names communicated. Each newsletter included items written by nonspecialists for nonspecialists. In 1945 alone, the newsletters included a review of an editorial in *The Washington Post* highlighting the new color-matching consciousness in sellers of menswear; a dissection of Helena Rubinstein's pseudoscientific "Color-Spectrograph," which highlighted different shades of reds for lipsticks that were "complementary" and "contrasting" and was plumped in advertisements as being based on science; news of a suit for brand infringement brought by the Yellow Cab Company against a competitor who painted its cabs orange, and the subsequent ruling by the court that "orange is a shade of yellow"; and the news that folks in Washington, D.C., had voted on color combos for their streetcars, the four combinations being "(1) burning-bush red and robins-egg blue; (2) maroon, with three 'matching tones' of tan; (3) peach-fuzz with maroon trimming; and (4) Kentucky blue-grass green and aluminum."[8] He took the increase of these soft-focus stories as a sign. The ISCC and National Bureau of Standards could strike while the iron was hot. And there was someone with deep ties to both organizations who knew the value of reach and had connections.

Godlove's letter to Merriam-Webster gets right down to brass tacks. "Knowing your policy of preparing for decennial revision in normal times, I am writing to say that I would be glad to revise, at no cost to you except stenographic charges and stationery, the terms and color names."[9]

Red Tape

Godlove's letter to Merriam-Webster in April 1945, offering to revise all the color definitions for the new edition of the *New International*, makes a lot of assumptions about Merriam-Webster's operations and the leeway that he'd get as a consultant. The PS on his letter finishes with a very Priest-like "My only requirement is that I be not hurried."[1] Surely Merriam-Webster would be happy to follow his schedule, since he was the expert.

But things at Merriam-Webster had changed, too. There had been a changing of the guard in 1934—a normal occurrence after the production of an unabridged, when everyone is burned out and desires nothing more than a job approximately five hundred miles away from lexicography. Of the thirty-nine editors and assistants listed on the masthead of the *Second* (including Carhart), only fourteen remained after publication. The editorial roster was riddled with holes, but as is often the case, three editors who had done the bulk of the general defining and revision for the *Second* rose up the ranks to become America's Next Top Drudge. Lucius Holt, a Yalie and a professor at the U.S. Military Academy who had also been a revising editor on the 1909 *New International Dictionary*, became the managing editor in 1934 and harrumphed in the great leather chair for twelve years before sighing retirement-ward. John Bethel, a Harvard man and former teacher at the Buffalo State Teachers College, became the general editor in 1935. And then there was Edward Oakes, who with merely a master's degree from Harvard didn't get a titled position, but who was Bethel's and Holt's right-hand man: the one who put all the ideas into action.

The three men were all roughly the same age but inhabited different eras. Holt never left 1909: Professorial, formal, and distant, he had the bearing of someone who wore an invisible celluloid collar all day, every day, even in the bath. Bethel was like the quiet, reluctant hero of a Stephen Crane short story: serious, unsmiling, touched with a little ennui. In a 1936 memo, a muckety-muck at Merriam asked the editor in chief to write some letters of professional introduction for Bethel, because he "is so modest and retiring, and always side-steps any social functions."* And Oakes is 1955 through and through: genial, smart, not afraid to do things his own way. In an interdepartmental memo from the staff librarian asking if anyone had copies of two books at their desks, Bethel, the first to get the memo, wrote "I have neither of these" under the query and signed with his initials and the date. Every other editor on the floor responded by lining up five ditto marks under Bethel's response—"″ ″ ″ ″ ″"—and signing/dating as appropriate. The march of marks proceeds down the rest of the page like a phalanx of mice prints in snow, and is interrupted only once by Oakes, who wrote, out of alignment with the others, a breezy "ditto."[2]

Oakes was a native son: born in North Adams, Massachusetts, graduate of Williams College (valedictorian, class of 1916; honorary master's degree, 1932) and the aforementioned Harvard. He did

* Of course he did: *He was a lexicographer,* a person whose natural reaction to a party is first silent panic, then hives, then anaphylactic shock.

a teaching stint at the University of Michigan, and then at Union College in New York, but came back to Massachusetts to work at Harvard's Widener Library until joining the ranks of the drudges in Springfield in 1927. He had a bout with polio that left him in a wheelchair—when the G. & C. Merriam Company was building its new offices in 1933, it had to put in an elevator for him since the editors worked on the second floor*—but contrary to what everyone around him might have assumed, it didn't dampen his spirits much at all. The citation given to him by Williams when it conferred its honorary master's degree on him called him "an athlete of the mind and spirit, quietly and cheerfully mastering fierce fate by indomitable will and brilliant scholarship."[3] He didn't have Bethel's shy seriousness or Holt's military bona fides, but he was smart and persistent—two qualities that, along with some long suffering, make for an excellent lexicographer.

After 1934, the staff stuttered along with other dictionary projects—editions of student dictionaries, a thesaurus, the *Collegiate*—but the *Third* was always on the horizon, and Oakes, in particular, was charged with much of the preliminary work for the *Third*. By the spring of 1938, he was in charge of the New Words supplement that would eventually be slapped into new printings of the *Second* and would serve as the first batches of new vocab for the *Third*, in an ideal world. In 1939, he was put in charge of devising the nascent reading program for the *Third*—evaluating periodicals, coming up with an analysis of how to best proceed, and working on getting the remaining sources that have already been read for new words typed up and filed away on index cards. Nothing too exciting there; this sort of work was done right before every major dictionary revision began. But what made Oakes's new program

* The elevator is still there, now used for freight, and probably hasn't been updated since the building was built. It's a loud, dark, shuddering contraption, and it makes me sad to think that this is how Oakes began and ended every workday.

unique was that it was ongoing and relied only on the trained in-house staff instead of volunteer readers. Oakes's program became standard operating procedure at Merriam-Webster for systematically tracking language from its inception onward; it's still done, and likely will be, as is the way with lexicography, until the inevitable implosion of the universe.

It seems like a small innovation, not one worth even mentioning, but it was evidence that Oakes understood that the world was changing, and lexicography, which had been strolling behind history at a scholarly meander, needed to start trotting. Between 1909, when the first *New International* was issued, and 1934, when the *Second* came out, the blender, the traffic light, the polygraph, the garage door (and garage-door opener), the grocery store, the crystal oscillator, the iron lung, the electric guitar, the radio telescope, autopilot, Freon, tampons, the cyclotron, and toggle light switches had all been invented. In that twenty-five-year spread, we discovered the structure of the atom and the nature of the Milky Way; we found a tenth planet, Schrödinger talked about undead cats, and Einstein talked about the bends in the universe. Housewives came to expect all the modern amenities, like refrigerators, pop-up toasters, and a radio in the family automobile. Here in the United States (some) women got the vote and the temperance movement got (some) Prohibition. We lived through the Jazz Age and the Great Depression, the Ku Klux Klan and Marcus Garvey, jukeboxes and talkies, the Scopes Monkey Trial and "Rhapsody in Blue," Babe Ruth (baseball player) and Baby Ruth (candy bar, no relation). Between 1909 and 1934, the words "poikilotherm," "angioplasty," "genome," "Nazi," "robot," "teenager," "schizophrenia," "evacuee," and "chemical weapon" first appeared in print—just a handful of the tens of thousands of new vocab words coming in to describe our brave new world.*

And yet the chasm between those who could readily navigate the scientific age and those who couldn't was already wide enough. "One of the greatest troubles with us scientists," writes the scientist Willis R. Whitney in 1918, "is our exclusiveness and aloofness. . . . The individual desire to see America advance in knowledge and proper activities is now everywhere. Even a dub can enjoy it," he marvels.†[4] And the dubs of the world were taking notice. Magazines like *Popular Mechanics*‡ and *Popular Science* began to become more popular in all senses of the word. *Popular Science* shifted gears in the 1910s away from heavily technical articles to easy-to-read articles on modern inventions, breaking news about futuristic tech, and interviews with the leading innovators of the day. In 1918, it boasted 150,000 copies in circulation, both in subscribers and in newsstand sales; in just ten years, its subscriber base alone was 350,000 people.[5] People wanted to be in the know; they just didn't want to have to get a PhD in physics first.

Oakes had his eye on the future, and in early 1944, when the future finally crested the horizon and began casting its long, dictionary-shaped shadow over the office, he was ready. While the in-house staff was being literally decimated (sense 1) by the war effort and the slow smolder of lexicographical burnout, Bethel, Holt, and Oakes began planning for the *Third* in earnest. Well, Oakes did: The first two asked Oakes, who was familiar with the content being gathered by readers, to do analysis of the *Second*

* A Shakespearean phrase that was given new life in 1932, thanks to Aldous Huxley and his dystopian novel of the same name.

† "Dub" is defined in *Webster's Second* as "*Sports, Games, etc.* A bungling, unskillful player. *Slang*." How rude.

‡ For which one John Percival Bethel wrote early in his career, before he became an editor at Merriam-Webster.

and estimate what kind of work would need to be done for a June 1954 publishing date for the *Third*. (Oakes, after all, was safe from the draft.) His goal was to get a sense as to what was in the *Second* by volume, and how much of it could—or should—be carried over into the *Third*, given how quickly language and society were changing.

His analysis, all thirty-eight pages of it, was impressive in its scope. He included careful markups of pages from the *Second* to note how much new vocabulary had been added between 1909 and 1934; what sort of stuff could be dumped; what stuff was likely to need expanding. He included supplements that broke down, by column inch and page, how much of each type of "non-dictionary" stuff was in the *Second* and how much space could be saved by dropping it. He wrote long memos on how to handle the encyclopedic info that people expected in their dictionaries but that could be dumped or cut down dramatically for space—trademark and trade names, lists of mythological characters, archaic terms. He asked another editor to check his own math and included that editor's work sheets with his.

His analysis, presented in an editorial meeting in October 1944, was also unsparing. The *Second* couldn't be expanded: They were already straining the binding technology of the day, and splitting the new, expanded *Third* into two volumes was the sales kiss of death. Content would have to be cut, and he proposed cutting the stuff that the sales folks (as well as Neilson, still the editor in chief, and Robert Munroe, the new president of Merriam-Webster) loved to point out after thwacking open their sample of the *Second:* the biographical and geographic sections, the article on the history of English, the forms of address, as much of the lengthy pronunciation guide as possible, the lists of battles, names, wars, fictional characters, ships. The new *Third* had to be a "pure dictionary," as he wrote.[6] But there was yet another catch: There was absolutely no way in hell that they could undertake a full *A–Z,* every-entry revision of the *Second* in ten years with the current (dwindling) staff. He had some ideas, but the fact remained: To bushwhack their way through 550,000 entries, they were going to need more people.

The editorial meeting minutes reflect significant dithering on the part of the head honchos in the room. Neilson thought that the *Third* should be bigger than the *Second,* and sales staff be damned if they can't manage to sell a two-volume version of America's Foremost Font of All Human Knowledge &c. Robert Munroe, a former member of the aforementioned damned sales staff, vociferously defended the inclusion of advertising pages in the *Third:* What's more modern, more forward-looking, than advertising? Bethel and Holt did not pitch in. The next year, Holt would retire; Bethel also had his eye cast toward retirement and was therefore trying very hard not to get promoted to the top editorial position of the company. Oakes never wavered in his view that the *Third* should be a "pure dictionary," though that 1954 pub date was impossible. He didn't push the point, but he was a stolid, bristle-brushed presence in the conference room. He threw the wheel locks on his wheelchair into position and dug in for the long haul.

With everything going on, it would be very easy for Godlove's April 1945 letter to be set aside, accidentally filed in a manila folder in the basement, ignored; they had nothing to say to him, really. Considering the paucity of responses Godlove got from Carhart the last time round, and the relative editorial isolation in which Godlove worked, there's a surprising amount of ink spilled in the Merriam office in trying to figure out how to respond.

Oakes became the de facto respondent, although because Bethel had the fancier title—associate editor trumps assistant editor—it was his name that ended up on the letter sent out. Oakes was certainly one of the most qualified editors in the office to begin the consultant wrangling; he was also, frankly, one of the more outgoing and *normal-seeming* editors, and so was a natural to take up overseeing the outside consultants. Bethel was too reserved and cerebral; Holt was too hidebound by the formality of the past. Oakes was the complete package: a seasoned editor who wasn't afraid of what lexicography could be, and who also could hold a conversation without becoming suddenly awash in flop sweat. But that's not to say that Oakes approached the correspondence with anything approximating enthusiasm. He sent his drafted response along to other editors for comments, and the note that accompanied it began, "This draft may be too callow as is, but it conveys something of our frustration."[7]

There are many sources for Oakes's frustration. The first is that Oakes knew from his *Second* days that Godlove essentially dropped out of view in 1933 after Carhart's sudden death, and yet here he sweeps in without preamble,* pretending to understand their publishing schedule and sounding very full of himself as he offers to revise definitions that no one has time to even consider revising.

He was also irritated with the supporting information Godlove gave in his letter to persuade Merriam to get going on the color defining again. Godlove teased that a new NBS-sponsored be-all and end-all of color standardization was just around the corner. Oakes had the correspondence file right there: Wasn't he saying the same thing thirteen years ago? To explain just how much had changed in the color-science world since 1934, Godlove sent along a list of publications for Oakes to review. There's the pinprick, the paper cut to the ego: Oakes was a scholar, the head of the read-

* Godlove had addressed his letter to Knott; maybe he assumed that there wouldn't be a need for a preamble, since Knott had been around during the *Second* and knew about Godlove's widowerhood.

ing program, someone who had done his damned homework. His note to Bethel ticked off, book by paper by symposium collection, how he had already read each of Godlove's supporting papers, and he didn't see any evidence of any consensus between any of these groups that Godlove mentioned on any sort of color standard.

Oakes was also completely underwhelmed by the work Godlove had done in the *Second*. The color plates that Godlove had put together were supposed to be the sine qua non of color and instead were *que será, será*. Priest's demand for more plates went nowhere; Godlove was given a princely two full-color plates to work with, and Oakes felt he had squandered them. There is the full-color painting of the visible spectrum, which Priest had insisted on, but which was functionally useless to the viewer. There were color samples; one could assume they would be orderly, or sensible, or explained. They were none of those things. Half of them look as though the photographer sneezed sometime between setting the chips down in orderly rows and pressing the shutter button. The remainder are in rows, but the chips themselves are in different sizes, different collections of colors, and unevenly labeled.

Godlove had made the mistake of thinking that these color charts were meant to be lessons rather than the teacher's key to the test. This means that the color chips reproduced on the plates don't have a one-to-one correspondence to the entries in the *Second*. The kind-of orderly rows of colors are meant to teach the viewer different things about color and color perception, like how contrasting colors affect each other, or how colors of the same saturation but different hue hold together. Lovely—now, if only the reader cared about those things instead of finding out what color, exactly, the "yellowish yellow-green" of "holly green" is. (No joy there: There's no chip labeled "holly green" anywhere in that array.) And while Godlove did provide a key of color names for the chips shown, that key is, like the charts themselves, inconsistent and contradictory. Set aside for the moment that there are hundreds and hundreds of colors defined in the *Second* which have no chip on this page to orient the reader. Instead, let's survey the chips as they are. There is no color chip for "orange," though there is one for "pink"; there's a chip for "deep chrome yellow," but not one for "medium chrome yellow" or "light chrome yellow." "Cherry" is chipped twice, the same color taking up a precious extra half-inch square of paper where maybe "holly green" or "light chrome yellow" could have gone. "Henna" has three chips to choose from, all of them a slightly different orange—which one is "henna," the reader asks, and why are these other two colors also named "henna"? And to finish it all off, some of the color chips are unnamed—which is not to say that they aren't listed in the key. They are: They are called "unnamed." You might wonder why in God's holly-green earth they were listed and photographed at all. No one knows, not even the poor Merriam-Webster editor with the salt-and-pepper high-and-tight who was tasked with answering the correspondence about why the color charts look like that and now had to explain that to the charts' original creator.

And the definitions themselves were insupportable. Godlove's work came in so late that there was no time to make sure that all the definitions he turned in held to the agonizingly derived formula for them: "a color, red/yellow/green/blue in hue, of high/medium/low saturation and high/medium/low brilliance." Some did; some, like "admiral," which Godlove highlighted as needing complete revision, were practically unchanged from earlier definitions ("admiral" had been defined as "= logwood" in earlier editions; in the *Second,* it ended up being defined as "the color logwood"). A handful of colors were dumped, but most were kept, despite Godlove's insistence that he would clear out the needless entries. Even the ones that had a comparative color attached to them were inconsistent: If you compare the color "cracker" to "aloma," that assumes that your reader knows what the color "aloma" is—and Oakes was led to believe that, no, the reader did not know such things.

And that, really, gets to Oakes's primary frustration with Godlove's breezy letter: He was a lexicographer and Godlove wasn't. It was Oakes and not Godlove who had had to live with and defend all those half-usable color definitions and that long thesis on color in the *Second*. He had given them significant side-eye as he planned the *Third:* They were exactly the sorts of academic inanities that he was increasingly convinced would have to be dropped. He felt, in his deep waters, the new dictionary for a new age was likely going to be a vastly different book from the *Second,* so Godlove pushing this color standard as the new way to think about color, from the top down, peeved Oakes. "As a proposed codification [the ISCC-NBS method] concerns us," Oakes goes on, "but it does not become imperative for incorporation in our general dictionary until there are unmistakable signs of adoption."[8]

This throwaway sentence is a seismic shift in the approach to the color definitions. The work of the consultant in any encyclopedic dictionary is to be The Expert, the one deciding what is worth entering and why, the one writing the definitions for their subject area. During the *Second,* The Experts were the guys outside the office, not the drudges therein: Whatever they turned in was, in an ideal world, just going to get a quick-and-dirty copyedit so that their wisdom and expertise could shine off the page. Oakes's reminder that the general dictionary was really about general language moved the power away from the consultant.

The response that Oakes drafted for Bethel did not mince words. He ticked off, one by one, the complaints they had received: Correspondents in specialized fields complained that the colors used in their industry, like leather dyeing or millinery, weren't in the *Second,* but there are plenty of other color names that no one has ever heard of littering the *A–Z;* other correspondents were irate that the descriptions for "simple" colors like "cardinal," "slate," and "ruby" didn't match real-life cardinals, slates, and rubies. "In effect," his draft read, "the excellent schematic def[inition]s that you prepared for W2 [*Webster's Second*] have proved serviceable only to the select few."[9] It was so bad that Oakes, the word guy, began to question whether words were up to the task. His draft continues,

> In fact, we have considered the advisability of omitting all defs of individual colors and entering only a cross reference "See color chart page —," where we should reproduce 300 or so colors.[10]

It was a good idea—radical, and it would require a lot of buy-in from the board of directors and people accustomed to buying their dictionary for all the words, not the pictures. Oakes made a note to himself to raise it in an editorial board meeting. But the question of what to do with the color definitions, and any sort of further conversation with Godlove, were quickly back-burnered when the editor in chief, Neilson, died in 1946. Godlove's letters languished. Everything would have to be sent back to the drawing board; a new editor in chief would demand it.

The Gray Head

It was a pretty good gig to be the editor in chief of a large dictionary around the turn of the twentieth century—mostly because there was actually very little editorial work required. The editor in chief was a figurehead, a recognizable scholar who would be able to lend some gravitas and legitimacy to this expensive venture. Neilson had been perfect for the job: president of a women's college with a scholarly tour as professor across a bunch of the Ivies; a well-known Shakespeare scholar who edited two of the core collections of the Bard's works; a fellow of the American Academy of Arts and Sciences and the American Philosophical Society. The Merriam board had big shoes to fill, but they were certain that they'd have plenty of feet to pick from.

Before his death, Neilson convened a meeting with Munroe, Holt, and Bethel to talk about who was going to succeed the soon-retiring Holt as general editor. About twenty names were kicked around; they include some of the superstars of twentieth-century linguistics and lexicography (Clarence Barnhart, Allen Walker Read, Hans Kurath, Charles Fries, and Leonard Bloomfield, notably) as well as some former Merriam editors and a clutch of English language professors. But nearly all twenty were tagged with major concerns. Too old, too focused on literature and not language, too soft in the scholarship department, too difficult to work with (a charge leveled at a surprising number of otherwise great candidates, further proving that dictionary people are definitely not people people). One candidate was rejected out of hand for liking Nazis too much; another because his short study at Oxford had turned him into an affected, humorless fud. Neilson suggested they just hire Alexander Murray Fowler, a grammarian at Harvard, but the board dragged their feet. But they had dallied themselves into a crisis: By January 1946, the seventy-six-year-old Neilson had fallen ill and was declining rapidly, and it was clear that they would quickly have to fill the top *two* editorial positions. Merriam hastily hired Fowler as general editor in January 1946, just weeks before Neilson's death, and immediately shunted him into acting as EIC. But Fowler returned to academia after just eight months; the editorial board was quickly reconvened and started another search. Oakes was invited to weigh in, since he had done the vast majority of the cogitating about the *Third* and would likely have opinions about the man to lead it; the board also called back Percy Long, a retired editor who could provide Neilson-ish guidance. Both editors brought a long history of the company with them; if anyone understood the importance of maintaining the Merriam legacy, it'd be these two.

Long got right down to it. He opened the initial search meeting with, "A primary choice lies between a linguistic scholar and a scientist. The humanities are passing, displaced by science and social science."[1]

If his fellow editors squirmed, it was more likely at his declaration that the humanities were dead than that a scientist might be a good option as editor in chief. This had been tentatively suggested to Long in the confidential letter Munroe sent inviting him to join the search committee. Munroe felt they had pretty well exhausted the crop of good English professors; besides, he had sales on his mind. "For its advertising value and its assistance to our commercial effort as a whole, the name of a prominent scientist might very well be of far greater importance to us for the next large dictionary than the name of an equally prominent linguist," he wrote. "Moreover, the 'man in the street' is more likely to know of a prominent scientist than of a prominent linguist."[2]

He had a point. America was one scant year out from the end of World War II, which had truly been a war won not on the size of infantry nor the courage of the soldier but on the smarts of the scientist. In the 1930s, European science all but collapsed: The Great Depression was a worldwide event, and Germany, the seat of modern physics and chemistry, was hit particularly hard. The country had been in political turmoil since the end of World War I, with democratic parties, communists, and nationalists warring (sometimes literally) for power, and the destabilization of the economy didn't help matters. Neither did the fact that Germany was still paying war reparations to the Allies. The Nazi Party began gaining parliamentary ground in the 1930s—Hitler's promise to restore law and order to the country and to bring it out of an economic slump, despite presenting no actual plan nor any evidence that the Nazi Party was able to do such a thing, was beguiling—and in 1933 the nationalist, anti-Semitic, anti-Bolshevik, anti–Allied forces Nazi Party came to power. That same year, the Reich began burning books and passed the Law for the Restoration of the Professional Civil Service, which allowed for the firing of any civil servant who opposed the Nazi Party or who had one Jewish grandparent. Nearly 15 percent of the professors in Germany were fired; scientists began leaving the country in droves. America and Britain were happy to welcome them with open, research-happy, fully funded arms. The 1936 List of Displaced German Scholars, put together by a German physician in exile and intended to help dismissed academics find work outside the Reich, includes Albert Einstein (whom everyone knows), Max Born (who helped give us quantum physics), Otto Frisch and his aunt Lise Meitner (who together helped discover nuclear fission), and Erwin Schrödinger (he of the undead cat).[3]

Once the Allies realized that Germany still had a number of scientists of quality working for the Reich—and once the Reich realized where its dismissed scientists had gone—it changed the tenor of the tensions in Europe. Military operations no longer just included the seizing of territory: The research capabilities and operations of each side were suddenly of tactical importance. Each side harnessed scientific know-how to improve its radios, planes, tanks, amphibious vehicles, subs, radar, encrypted communications, weapons. Wars were no longer fought by infantry, cavalry, navy alone: After World War I, it was assumed that science marched alongside troops in every traditional theater of warfare. But just as chemical weapons introduced a new level of scientific threat in World War I, Otto Hahn's experiments confirming Frisch and Meitner's theory of nuclear fission in 1938 opened up a new, global vista of potential domination and destruction in the latest world war. A nuclear-powered bomb, a superweapon, had unmeasured destructive capabilities that would require almost no sacrifice of men on the aggressor's part, but could effectively and immediately wipe out areas many times the size of London or Dresden. The Manhattan Project in the United States started in response to the discovery in 1939; around the same time, a parallel program called the *Uranverein* (Uranium Club) began in Germany. None of this was the work of the bespectacled English profs of Harvard. We had fully entered, whether Merriam liked it or not, a scientific age.

A variety of scientists were considered for the editor in chief position, notably James B. Conant and Harold Clayton Urey, both involved in the Manhattan Project. But in true dictionary fashion, their academic credentials were considered above anything else: Urey's Nobel Prize in Chemistry went unmentioned, but his academic CV was picked over meticulously. All the recommendations were brought confidentially to the board, and the actual soliciting was left to Munroe. (Oakes sent his own thoughts to Bethel, the notes in his distinctive purple ink stapled to the in-house copies of the letters of inquiry: "Any other chemist than Conant I should contest as editor in chief as too far divorced from matters of living language";[4] "Urey might allow us to dispense with physicists";[5] and so on.) Oakes contributed by writing out a job description. The new editor in chief should easily be able

> to speak with authority when the issue is in his own field, to sit back as arbitrator when the issue is out of his field, as an experienced administrative officer to quiz editors and business representatives on practicability. To act as a brake on personal idiosyncrasies, excessive zeal, unnecessary refinement of the editorial material, and anything else that threatens to prove detrimental to the intelligent consultant of the revised dictionary.[6]

In short, they were looking for Jesus Christ, PhD, MBA.

Munroe was having a hard time selling the position, though. The lure of underpaid, nebbish work in service of Mother English and your last name on the masthead was not as shiny as Merriam thought it would be. As he struck out, he began to reconsider the tack: These scientists all wanted *money,* and they didn't seem all that interested in English.* Interviews were held and offers made,

* He confidentially reached out to one of the science consultants from the *Second* to ask him for recommendations. Harold Bender definitely had a recommendation: ditch the idea of hiring a scientist. "It would be a sop to current popular interest,"

but no one bit; back one more time to the (dusty, dilapidated, unappealing) drawing board.[7]

Much has been written about the process of writing dictionary definitions, but not much has been written about the sort of in-house preparations that take place before any definitions can be made. This is generally because these preparations are, to 99.998 percent of the general populace, smotheringly boring. To lexicographers, though, this is the air we breathe and the water in which we swim. It is also the primary—and, in some cases, the only—editorial responsibility of an editor in chief.

When a lexicographer speaks of "editorial style," they aren't referring to the type of dictionary they're writing: This one's a children's dictionary, this one's a fancy dictionary, this one's a cheapie paperback. They are referring to the nuts and bolts of each part of a dictionary entry. What's the order of information given? How are the syllable breaks in the word shown? What sort of information is needed in the etymology? How is the etymology set off from the rest of the entry? Are we abbreviating the parts of speech or no? How do we mark the division between the sense number and the start of the definition proper? How many spaces are between the sense number and the definition?

The answers to all of these questions are bound by space and time, and not in the cosmic sense. The *Third* was going to be a single-volume dictionary, which meant that there are only so many pages in play, and only so much space per page to work with. The actual printing and binding technology wasn't necessarily the

---

he wrote in response to Munroe's letter, "but would, I think, serve no other good purpose. The dictionary is a recordation of the language not of science, much as we want to keep up with the latter, which is only a part. It is undue emphasis that I should fear, as well as, let us say, not the best editing."

issue—though the *Second* certainly was pushing it. It was trying to figure out how much people would spend for this new unabridged dictionary, taking that number, and prognosticating from there. The marketing folks and the president will take a look at past sales and comparable books on the market, and maybe hire a firm to do a little field research, and determine that people will spend exactly $45 on this dictionary—not $46, and not $44, since $44 sounds cheap enough that you wonder about the quality of this Most Authoritative Doorstop you're about to buy, but $46, whoa, Nelly, what's it printed in, gold?? Then the president takes that number to the managing editor, who will look at old contracts with the compositor, the typesetter, the printer, and the bindery to figure out what their costs per page will be, plus, say, a 20 percent margin (because those costs will have magically gone up since the last printing of this big boondoggle). Then the editorial board starts looking at how to cut costs without cutting corners: Can we do this with a skeleton staff? Can we hire low-paid college students? Can we do away with color plates? Can we use the same Forms of Address back matter from last time? How many lines will we save if we shorten the "from" in every etymology to "fr."? How many lines will we save if we shorten the "from" in every etymology to "fr" without the period? The goal is to get as much new stuff in the limited number of pages allotted to you. And from there, the editor in chief begins to build the style guide. It's both macro—no gazetteer in the back of the book—and micro: Abbreviating "especially" to "esp." in every definition in which it appears will save fifteen pages; therefore, the new house style is to abbreviate "especially" as "esp." in definitions.

This sort of fiddly finagling was Oakes's domain. He had already been training new editors as they came in, particularly when Bethel was too busy to do it (which, given Holt's departure, seemed to be often), and as the one who had done the analysis on the *Second,* he was in a good position to begin setting style for the

Third. The Merriam archives are littered with memos of his from 1946 onward suggesting changes in preparation for the next Big Book: what should stay in (new zoological terms, defined according to a paradigm he set up), what should go (mythological names, the history of literature, dialect words, words marked "erroneous," "rare," or "archaic," obsolete words, proper names, names of people groups, proper adjectives, and about three dozen more kinds of words). In a 1947 meeting to discuss the *Third,* Long said he thought that the way that the first *New International* had been done, relying more on in-house editors to write definitions, was vastly superior to the way that the *Second* had been done using all those consultants, and that if Merriam could just hire editors in certain fields, this would be solved. Oakes tried to nail that Jell-O to the wall—How many specialties? Which specialties? How many subspecialties?—to no avail. He carped in a note to Bethel, "I'm fed up with Percy's complacent attitude that the solid features of [the *Second*] were his, that when his advice was not followed the Book went wrong, that Knott and Baker were crude philistines."[8] Oakes was preparing for the future, and all this farting around in the past wasn't going to move the book ahead.

The lack of leadership at the top was also not going to move the book ahead. Bethel and Oakes did their best to keep things running, but without an editor in chief all they could do was tread water. In 1950, the company hired William Freeman Twaddell, a professor of German from Brown University, as editor in chief; like Fowler, he would stay less than a year before fleeing back into the loving arms of academia. It was Oakes who picked up the pieces when Fowler and then Twaddell decided, right after taking a look at this unwieldy omnium-gatherum, to hightail it out of Springfield. A glance at the process memos that were prepared* for the

* By Oakes, obviously.

third incarnation of the EIC tells a clear story in an easily tickable list. Editors responsible for gathering and collating information in advance of the EIC's arrival: Bethel and Oakes. Editors responsible for editorial conferences with the EIC: Bethel and Oakes. Editors responsible for sending out editorial questionnaires: Bethel, Oakes, another newly hired junior editor. Editors responsible for expansion of the reading program: Oakes and Anne Driscoll. Determination of editorial policy for the *Third:* Twaddell, Bethel, Oakes. Recruiting and training new editorial staff: Bethel and Oakes.

All this extra work didn't go unnoticed—at least, not by the other editors. Bethel and Holt sent Munroe a memo noting that every resignation by a new (another) EIC "makes Mr. Oakes' value additionally great," and insisting that the company up Oakes's pay so that he was at least making as much as this new snot-nosed linguist they had just hired, a Mr. Gove.[9] They also suggested he be promoted to associate editor, though he was essentially working as the managing editor. (Managing editor was a role they were saving for the new EIC's second-in-command; the new EIC would be the one to convey that title.) Munroe, no dummy, approved the raise and the new title. Oakes's letter of thanks was slightly guarded. "I value this expression of confidence on your part; it involves trust in my editorial judgment," he wrote, and then notes that he's told Holt that he really can't work any more than he's already working; he's clearly expecting the deluge of new responsibilities that will be dumped on him as an associate editor.[10] He wasn't disappointed. As Bethel focused on the production side of things and liaising with Munroe and the board of directors, and Holt packed it in to finally, finally retire, Oakes took on the day-to-day operations of the editorial floor. The files at Merriam-Webster are full of memos to and from Oakes about training new editors, readers, and typists; about the reading and citation collection program he was running; about editorial policies for the forthcoming *Third.*

His warning upon his promotion that he couldn't take on more work was tested again and again for years.

Twaddell had departed his short tenure as EIC in part because he thought he, the Outside Big Name, was unnecessary. "No one person, in scholarship, combines high reputation and balanced versatility as the Merriam-Webster editorial staff does," he wrote in a memo to Bethel, and presented a solution to the revolving carousel of EICs: Give the editorship to a Merriam editor who had "proved himself versatile, resourceful, energetic, able to imagine difficulties before they arise, able also to accept the fact that unforeseen difficulties will arise no matter how well the planning has been done."[11] The idea that an outside editor in chief would be able to run the editorial floor was laughable: It was a tightly knit group that needed daily supervision and guidance—well beyond the sort of work needed to get the *Third* going. He suggests that the new man in charge be conversant with the *Collegiate* line, and perhaps someone who can stave off the attacks that Merriam-Webster was out of date and certainly not progressive enough for a *modern* dictionary.

Bethel didn't need much convincing: The workload and strain of trying to keep the whole editorial floor together led to a decline in his health, and he wanted to retire. The board of directors would likely be happy to off-load some editorial expense—all those useless outside EICs—and get the *Third* going using someone in-house (though they balked at giving anyone in office the coveted title "editor in chief" and settled on "managing editor" in case some highly pedigreed unicorn happened to want the figurehead job at a later date). The choice was, given everything, obvious. "Confirming our conversation of June 25, you have been selected to have full responsibility, under the Editorial Board, for the editing and production of the next edition of the Merriam-Webster Unabridged Dictionary," the offer letter began. "We have full confidence that you will measure up to the responsibility of this most exacting work and that the Dictionary, when published, will rank in scholarship and accuracy with Dictionaries of the past. We expect that you will create a staff of efficient coworkers to somewhat lessen your personal responsibilities."[12] The offer was signed on June 27, 1951.

The new managing editor and acting editor in chief for the *Third* was not Edward Oakes but a new guy, the guy assisting Oakes in sending out editorial questionnaires and whose hiring had spurred Oakes's promotion and pay raise: Philip Babcock Gove.

The Greenhorn

Gove had been hired by Merriam-Webster in 1946 after he sent them an unsolicited application. He didn't ask if there were any positions currently open, but rather said that he'd like to work for the company in September of that year, if possible. Gove would have called that not presumptuous but direct. How else does one express their interest in a job?

His qualifications were laid out in one paragraph, not unlike the paragraphs that the editorial board had put together for all the other editor in chief candidates. PhD from Columbia in English and comparative literature; one-year research fellowship in England from 1939 to 1940 (which began with the evacuation of children from London and a declaration of war, and ended early when the Goves were warned to be on the last ship bound for the United States before the German U-boats made transatlantic travel impossible); English instructor at Rice University and New York University; lieutenant commander in the U.S. Navy, where he had been in active service on the West Coast from 1942 to 1945; commander in the U.S. Naval Reserves; aged forty-three; married; three children. "I am not a linguist and have no claim on being a lexicographer," he writes, doing his damnedest to dissuade a bunch of lexicographers and linguists from hiring him, "but have done considerable research on 17th- and 18th-century dictionaries, have about as much knowledge, as an outsider could acquire, I believe, of how dictionaries are made, and know something about printing and preparing Mss. for the printer."[1] It worked: He joined the staff as an editor and was trained by Oakes.

Gove has been called many things by biographers, co-workers, and critics, but all seem to agree on a few character traits. Efficiency was practically his religion, and he grew immediately irritated with any sort of time-wasting foofaraw. This included being collegial. Gabbing at the watercooler about sports—what was the *point*? He was detailed and meticulous in all things, and particularly his use of words: One reason why he didn't pursue a career in the navy was that he couldn't deal with the dizzying circumlocutions of government writing, and he was swift to judge other people's floppy use of language as evidence of laziness. Diplomacy was not his strong suit: He was passed over for a promotion in the navy because his commanding officer had written that while Gove's performance was satisfactory, "he has rather definite convictions and at times is not tactful."[2] A sharp mind, certainly, but encased in the hardest of heads.

Oakes, as his supervisor, swiftly learned that his new young charge could be a real asshole when he thought someone was wrong and he was in the right. As the functional head of everything, Oakes was in charge of the revision and changes made to the *Second,* which meant he was the one in charge of organizing new entries that needed to be added to each new printing. In 1947, the armed forces of the United States gained a new member (the air force); Gove, being a navy man, knew about the change, and so wrote a tentative definition for "air force." Oakes didn't think all that much of it. He wrote to the Department of the Air Force asking them to provide the company with a decent generalized definition of their agency for comparison, then revised as was his right and his duty. Oakes's revision was in the style of the *Second:* encyclopedic, with decent chunks of the air force's proposed definition left intact. His revision reads, "A military aviation established; esp U.S., a component of the armed forces constituted July 26, 1947, having jurisdiction over all aviation units required for air defense of the United States, strategic and land-based tactical air warfare, and all other air operations not organic to other military services."[3] The three-by-five-inch defining slip was attached to the production changes sheet, where Gove came across it and the supporting letter from the publications branch of the air force.

He made his feelings known by writing a word-by-word refutation of what the chief of the publications branch of the Department of the Air Force wrote. Never mind that Major Francis E. Skipp Jr., USAF, probably knew more about the air force than Gove did; Gove knew more about dictionaries. "Above definition (so-called) is one man's version of an insoluble tangle, unsupported by law and subject to the next shift in policy," he snipes. "Aside fr. fact that the PRO who relieves him in a few months will have a different version, his def. is full of weasel words which we should not take over into W34 [*Webster's Second*]." He goes on from there for two full pages, asking if "air defense" begins at ten feet off the ground or ten thousand feet off the ground, pointing out the logical fallacy of "land-based . . . air warfare," complaining about the placement of "strategic" and "tactical," and ending with, "My original def is still my choice."[4]

It's not unusual to draft an entry and be utterly dismayed and even hurt when it is inevitably revised by a more senior editor. You spent a lot of time on that draft, calculable in amounts of coffee and silence; it can be dispiriting to get the galleys back and see that the three lines you've ground your metaphorical gears over for hours have been slashed into a one-word synonym. This is part of what it means to grow up as a lexicographer, though. Having your definitions edited teaches you to let go of your nerdy little peccadilloes and pick your nerdy little battles. If a revision is actually, factually wrong, there is an established process for addressing that: You can go back to the revising editor and make a case as to why the revision is wrong. Otherwise, the editorial hierarchy reigns, and you must leave the definition in the hands of the more senior copy editor.

Gove absolutely did not do this. Instead of sending his remarks back to Oakes in defense of his original definition, as he should have, he commented on Oakes's defining slip, which was already clipped to the galley and on its way to the production desk, and noted on the back of that slip that the definition was wrong and should be replaced. Then he pulled his original definition, made one change to it using a production note so that it wouldn't go past Oakes again, and clipped it to the galley. Gove's slightly edited original was what went to print, not the revision made by a more senior editor. This was so far out of the norm that a production editor later left a note in the file explaining how that definition for "air force" came to be in the 1949 revision of the *Second* so that future editors would know (a) why there were two drafted definitions for "air force" in the file as opposed to one, and (b) that Philip Babcock Gove was going to get his way, so best get out of it.

Gove didn't lighten up in the following years, either. When he was made the managing editor of the *Third* in 1951, he immediately instituted a rule of silence on the editorial floor that came to mark the Merriam way: Talking was a distraction to both the editor who was diligently working and the editor who should be diligently working. He did not socialize with the other editors except at lunch, and then only in the company cafeteria located in the basement of the building. Other editors thought he was distant, standoffish, and unwilling to do the work of actually managing the editorial floor like a managing editor should. Even picayune matters would needle him. He argued extensively with one Merriam-Webster correspondent over the definition of "applejack" for nine years, and only stopped because he was technically retired—and had been for four of those nine years.

On the rare occasion when he was forced into a violation of his principles, he was a ruthless documenter of hurts and ills. The *Third,* when it finally did come out, famously did not include that Mother of All Curse Words, "fuck." The then-president of the company realized during the last stages of production that "fuck" had been entered and demanded that the entry be axed lest it injure sales; Gove argued and argued until the president told him to stuff it and do as he was told. He did. He resented it until the day he retired, and probably the day he died, and minuted his resentment in a handful of notes that documented every mention of the omission of "fuck" in every review of the *Third* he could dredge up for years afterward. He kept every one of those notes, scrawled on the standard-issue three-by-five-inch index cards that litter Merriam offices and filed his protests alphabetically, long after Merriam finally caved and entered "fuck" into its dictionaries.

Munroe and Bethel needed a take-charge sort of guy, and Gove had proved he was. They both assumed that his military precision and commitment to working long hours in strained circumstances meant that he would tolerate the intensity of turning this late boondoggle of a lexicography project into a best-selling dictionary sensation. Gove was no businessman, but he fully understood the direness of the situation. He had been put in charge of a project that was already halfway through its projected active production cycle and that had already devoured a quarter of the total editorial dollars over the previous sixteen years, and there was absolutely nothing to show for it: no sample definitions to show the stockholders, no big name on the title page to allay fears that this big money sink was going to eventually pay off. His promotion letter hinted at all of this. It is not full of lofty praise for work well done, no "the company is pleased to grant you," and so on. It is positively Govian: straight and painfully to the point. The memo gives him his promotion, his salary, and a short list of benefits and is capped off with "measure up" and "responsibility," "exacting" and "we expect."

Consequently, one of his first tasks as the managing editor was to do something more managerial and less editorial: He undertook a budget analysis of how much they had already spent on the *Third,* and in doing so, he requested copies of all memos, notes, and correspondence related to the creation of the *Third* (including copies of confidential memos sent to outsiders asking them for recommendations on a new editor in chief—records are records). That's how he came into possession of all of Oakes's notes on the *Third,* which he promptly ingested and which became, then, his notes on the *Third*.

Armed with this evaluation, he made a thorny nuisance of himself from the jump. In his first editorial board meeting after taking charge, he countered Munroe's sweeping introduction about how this kickoff meeting to begin work on the *Third* was "one of the most important meetings this Company has ever had which I have attended" by immediately pissing on Munroe's parade.[5] The first on-record statement Gove made as the managing editor was that this new dictionary is already a mess. He started with a lexical analysis of each part of the title of the book, noting how squishy each constituent of *Webster's New International Dictionary* actually is in the bright daylight of actual English usage, and finished with what I'm sure he thought was a statement that smoothed things over: "I point [this] out simply to show that the problem is larger than the Editorial Board or than any single member of the Editorial Department, and that it symbolizes compromises in the task we are undertaking."[6] Not exactly a ticker-tape moment, much to Munroe's chagrin.

The compromises, Gove explained, are many. The dictionaripedia was no longer feasible—from both a scholarly and a fiscal standpoint. He began ticking off whole categories of things that were extraneous in a modern dictionary: proper names, mythology, the gazetteer, the recondite technical terms from law, botany, chemistry. If this sounds familiar, it's because these were exactly the things that Oakes had ticked off as being dumpable back in 1946. Gove hadn't just read what Oakes had written on the subject but adopted it wholesale: Even the language Gove used to speak about this sort of "nonlexical" information (as the modern linguistic parlance had it) was lifted directly from Oakes's earlier preparations. When he called this information "museum stuff," he was quoting one of Oakes's 1946 memos to Bethel; when he rattled seemingly off the top of his head that eliminating entries for literary works and characters would save thirty-two pages, he was using Oakes's estimate.

But if Oakes was hoping that Gove's attitude to the encyclopedic material in the *Second* was a mark of fellow feeling, or of Gove championing the work that Oakes had done on the *Third* already, he was going to be bitterly disappointed. Gove didn't credit Oakes for the work already done, nor thank him for laying the groundwork that enabled a more junior editor (albeit one with a PhD) to step into the role. That was just the sort of sentimental flummery that wasted time, and they were, according to the original timeline, already four years behind schedule.

The other point of pique for Gove was the whole way that editorial decisions had been made for the *Second* and since the *Second*. The lengthy, academic back-and-forths that littered the editorial board minutes from the *Second* were inefficient and ridiculous. "I understand that the Editorial Board for *Second Edition* spent at least an hour discussing whether hot dog was to be entered," he marveled—and not in the good sort of way, either.[7] Oakes had been silent until then, but he spoke up. He agreed, but wanted to clarify: So defining questions were the remit of the editors, and business decisions were the remit of the editorial board?

Gove apparently willfully misunderstood Oakes: If Oakes was asking about what the editorial board should be doing, Gove focused on what the editorial board should *not* have been doing. Gove seized on defining memos, for some reason, as evidence of the sort of faffing around that the more senior editors should absolutely not be doing. These memos and the process that led to them—the editorial board setting defining policy, and Oakes in particular setting it down for the new lexicographers—were the points of inefficiency. Why was the editorial board fiddling around with "hot dog" and issuing these stupid memos on things like proper flower taxonomies when what was actually needed was more attention to workflow, efficiencies, clarity of purpose? "The Botanical Classification memo should have probably never been put in your hands," he sniped at Oakes, as collegial as buckshot at close range. "I could read it over a number of times and never know if it was any righter or wronger."[8]

According to the transcript, Oakes remained silent for most of the remainder of the meeting.

Munroe, former salesman, squirmed at hearing Gove claim all that lovely (salable) ancillary information wasn't *scholarly*. Merriam was known for its scholarship; the sales department had crafted entire booklets on this sort of scholarship;* readers would *expect* this sort of scholarship. Gove clarified. The *Second,* their model, was not

* Including a tremendous two-page spread that was tacked into the front of sales copies of the *Second* and was called "COD: AN EXCURSION."

just out of date but also out of mode with the scientific approach of the modern age—by which he meant current linguistic thought.

Linguistics had, while Merriam was Twaddelling their lives away, gained much more scientific cred than most folks at Merriam realized. During the nineteenth century, linguistics as a field went from being a domain primarily of guesswork to being a field in which the categorization and history of languages began to slowly become systematized. It began with the discovery (based on consonance between seemingly unrelated languages) that English, Latin, and Sanskrit were all likely descended from some long-distant ancestor and continued with the systematized reordering of how individual words evolved based on this new understanding of language (a field we today call etymology). During this time, the study of the historical ties that bound languages together went from being based on your own thoughts and opinions to becoming an orderly, logical, almost scientific sifting of languages into different groups.

Science—linguistic and otherwise—marches on: By the time Gove went off to Dartmouth for his undergraduate work, linguistics looked not just at the history of language but at its present. The focus was on grammar—the structure of language, how it was formed, what similarities there were between how different languages were structured, and, as technology improved, how the brain processed language. Linguists even went to war: Because the field had shifted emphases from history to structure, linguists were hired by the government to analyze, say, a bunch of encoded Soviet diplomatic documents which appeared innocuous but which were actually espionage reports that, once cracked by linguistic analysis, revealed a huge Soviet spy ring in the United States that passed information about the Manhattan Project back to Moscow. By the time Gove sat down in the conference room to school those present on how the *Third* should be handled, linguistics was well on its way to moving from "social science" to "hard science." This wasn't the nineteenth century anymore, and the *Third* needed to be written on new principles: *scientific* ones.

Take, for instance, the pronunciations. The field of phonology had continued to blossom well after Carhart left Germany and came back to Merriam to implement what he had learned. The *Second* had received some gentle mockery for its pronunciations, which did a fine job of recording the overstuffed speech of the Boston Brahmin but couldn't be bothered with any other part of the country. This needed to be remedied for science's sake. Pronunciations no longer would focus on the educated speech of New Englanders (who made up the bulk of Merriam's staff) but should capture the speech of the major American dialects in use. Further, Gove thought that the pronunciations should be given not in the modified alphabet that had been used for decades but in the International Phonetic Alphabet. The IPA was favored by linguists and academics for its ability to express, with laser-sharp accuracy, most of the mouthy sounds that a human could make. Not just sounds we associate with speech: If Munroe were to blow a raspberry at all this science-schmience talk, a phonologist could transcribe Munroe's spitty disdain into a set of IPA symbols quite easily.* IPA was also impossibly abstruse: If you think that pronunciation keys in modern dictionaries are just too simple, ðɛn jʊr ˈgɑnə ʤʌst əˈdɔr aɪ-pi-eɪ.† When the man in charge of educational sales worried that a scientific pronunciation key would make it difficult to sell the dictionary, since it would be far more complicated than

* [r̼̊], for the record.

† "Then you're gonna just adore IPA." Yes, "gonna" and not "going to" (which is ˈgoʊɪŋ tu), because IPA is intended to be used that specifically when transcribing speech, and because that's just the sort of slop-mouthed pabulum I go in for.

what's in the *Second,* Gove sneered, "We fell into the popular error of misusing the word scientific. Scientific doesn't mean the key is any more difficult; it only means that it is more accurate."[9]

Gove might have been unsparing, but at least he was also cranky about it. Shareholders be damned: If they wanted to get this dictionary done this century, they needed more editors, and they should bring in as many outside consultants as they could to write definitions for the more obscure fields so the in-house staff had time to shore up "linguistic abstractions." Yes, training new staff would take time; yes, herding consultants would require effort. But he saw no other choice.

They hired with abandon: When Gove took the helm, there were 26 Merriamites working on the editorial floor (including the typists and clerical staff); in one year, he had more than doubled the staff, and he kept hiring. But the pace of the work was blistering, and the editor in chief was a real piece of work. Editors could find easier work for better pay; further, they worked for a guy who prohibited them from talking to each other, rarely handed out a good word to anyone, and rabidly enforced the no-smoking policy in the building at a time when smoking in the office was not just acceptable but practically expected. In a booming postwar economy, Merriam began to hemorrhage people at an unprecedented rate: Of the 188 editorial staff hired after 1951 in the run-up to the *Third,* 81 of them left within four years, about half of those departing lexicographers actually left after two, and at one point 15 editors left in one six-month span. Gove never linked their departures to his managerial style or to his hiring choices. He liked pedigree and trusted academic credentials and knowledge more than he trusted proven work. One proofreader recalls getting the job not because he did well on the proofreading exam but because Gove asked him what the going rate for a first edition of Johnson's *Dictionary* was and the proofreader happened to know.[10] When hiring, Gove didn't think to warn people about what sort of pressure cooker they were slipping into. Very few people are well suited to the silent office, exacting precision, and mental torture that is unique to Govian lexicography; the guy doing the hiring, though, thought it was just de rigueur.

Because Gove shut himself in his office and literally didn't speak to other editors—he demanded that if a question needed to be asked of an editor, it be written down on a slip of paper and delivered to their inbox so that the editor's concentration and workflow wouldn't be interrupted*—complaints from the editorial staff had nowhere to go. Bethel, the most obvious choice, was busy coordinating the production of the other dictionaries and, as seemed to be practically required of every senior editor at Merriam, working himself into ill health. Of the three remaining associate editors who predated Gove and so might be good people to approach, one of them absolutely couldn't stand Gove, one of them was traveling to help get the Merriam pronunciations more in line with modern scholarship, and the last was Oakes. He had been the main point of contact for editors before Gove, and that wasn't changing after Gove either. With the extra workload of training the carousel of new editors, it was not a position he relished. He sent notes to Gove asking for specific editorial decisions so harried and mutinous editors didn't have to suffer Gove's direct supervisory condescension. Oakes alerted Gove to a request from the etymologists: What other sources were available to them if they couldn't find what they needed in the Merriam etymology files? Gove's response was acerbic and typical: "Why worry about them now?

* These slips of paper were called pinks because they were pink index cards, and they were still the primary mode of interoffice communication when I started at Merriam-Webster in 1998. Lexicographers *really* like their traditions.

What would ETY do with them if they were found except decide it is impossible to do anything with them? (a typical decision)."[11] Oakes had to then pretty up this information and swallow Gove's nastiness, presenting the answer in as bland and benign a way as possible. This he did dozens upon hundreds of times, the scapegoat for all of Gove's decisions.

In 1952, knowing how low morale was, Oakes held a meeting with the editors to help assure them that the company was on track and that the vast changes that were rolling out of Gove's offices weren't going to derail them.

The records of this meeting are extraordinary. The cover sheet reads,

> REPORTS OF OFFICE QUERIES
>
> These are brief notes on questions and criticisms put to E. F. Oakes recently.
>
> Sharp comments about personnel are not included.[12]

The questions are recorded anonymously, and they bear witness to the general sense of topsy-turvydom that many editors felt. Why the sudden preoccupation with linguistics, when the vast majority of dictionary users won't care about all that? How are we going to communicate between editorial sections if we're no longer allowed to talk to each other? Do other editors (read: Gove) resent when we ask questions or make suggestions? Who's on the editorial board? Who's making policy decisions? To whom should we direct our questions? One recorded query isn't a question but a comment: If you want us to throw in our lot with Merriam, you need to actually tell us what's happening and give us a reason to believe this company has a future.

Poor Oakes did his best, answering as diplomatically as he could, but some of his answers belie his fatigue. In response to the question about where editors should direct their questions, he notes that there used to be a suggestion box, but nothing really came of it. In response to the comment that the company hadn't done much to instill confidence in its editorial group regarding their future employment with the company, Oakes could only say that the delay between the *Second* and the *Third* was due to illness and death, and that they had to wait to appoint a managing editor before they could take active steps toward the *Third*. When he responded to the question about communication between editorial departments, the notes show that he's not thrilled with the status quo, either: "The present isolation of editors and readers is disturbing. . . . Morale could be built up by authorized interchange between departments of plans, methods, attitudes."

The passive voice in that last sentence speaks volumes. Even as Oakes attempted to get Gove to intervene in sticky situations where the managing editor would—and should—preside, Gove foisted the responsibility back on Oakes. Oakes would refer a question or concern to Gove (should we buy this very expensive book in a series for the reading program now, or wait until the whole series is out?), and Gove would merely sneer in response (what a stupid question—if we wait until the whole series has been released, nothing will ever get done, and "here we go back to 1920!").[13] The power-vacuum feedback loop was exhausting, and Oakes had plenty of other things to keep him busy. Like Carhart before him, about 202 other things.

Down in Black and White

Godlove had written a number of times throughout the late 1940s, mostly to let Merriam-Webster know he was ready and able, at any moment, to start work on the color defining for the new unabridged dictionary. Every letter was an update on the state of color science, along with an encouragement that Merriam wouldn't be sorry if they hired him again to handle color definitions.

Godlove chalked up the inadequacies of the color definitions in the *Second* to a different philosophy about color. Merriam had approached Priest, a guy at the NBS, to define colors; the whole point of the NBS at that time was to impose standards from the top down. Have a problem with finding a good color for your railroad signal lamps? Here's the spec. Making glass eyes and need to match glass iris colors so they look natural? We made you a handy chart. Want a list of color names and their definitions? Voilà. Don't argue, because the scientists have determined that this is what should be done. But that's changed, Godlove insists. The idea that dominated the Colorimetry Section during Priest's day—that scientific analysis is right and popular notions about anything color related were wrong—was waning.

The committee that came up with the ISCC-NBS method had come to realize that, just like a dictionary, a color standard couldn't be fixed. The Munsell nomenclature needed to be expanded and recalibrated if it was to remain useful; rather than junking it, many scientists saw the value in keeping this more commonsense translational system and expanding it. The ISCC-NBS group also decided that if they really wanted to make a standard, it needed to include sources that were baldly commercial in nature—design books, color systems used for interior design, fashion forecasts. If this was going to be a color standard to meet art, science, and industry, then they couldn't favor science over art, or some part of industry over others.

There was another thing that had shifted that affected the ISCC-NBS standard, too: language. The guiding principle had always been to use the most descriptive and easily understood words for the system—not the shortest, or even the most chromatically consistent—but a reevaluation is a reevaluation. For instance, "dusky" had more common meanings besides "blackish" or "dark": By the 1940s, it was used more often to describe dark skin than dark colors. Clearly, "dusky" as a *color* descriptor had fallen away. Other words had come into use—mostly through consumer fashion writing—that might fit better. All modifiers were being looked at again for currency.

Even the core color names needed an upgrade. One ingredient in the USP was described in the original as a "dark reddish brown." In Godlove's new notation, it was a "dark weak orange," and everything about that plain-language description shrieks bright red incomprehension. "Dark" and "weak" together are confusing enough, but how did we get from the USP's "brown" to "*orange*"?

"Dark weak orange" may well be systematically accurate, but it's a dark, weak pile of garbage to the average reader. What's wrong with "brown"? Godlove had come around to thinking that nothing was wrong with it.

If Priest had still been alive, that idea would have merited an all-caps, fully underlined, multipage screed costing him several typewriter ribbons' worth of ink. The initial problem with "brown," remember, is that it spanned too many color categories. You can have yellowish browns and greenish browns and reddish browns: This type of language would communicate to us slobs that "brown" is a discrete color that's different from red or green or yellow, as opposed to being a *subset* of red, green, yellow. But how valuable is scientific accuracy if people are confused by the scientifically accurate description? You're going to describe the color we call "chocolate" as "dark reddish yellowish orange"?

The Munsell re-notations also led to another linguistic problem. There are really three kinds of orange that pop to mind when you hear the symbol/word "orange." There is "unique" orange, or the orange that is exactly colorimetrically halfway between red and yellow; there is "typical" orange, which is the single color in a range of oranges that most people associate with the word "orange" alone and not a description like "carrot orange" or "safety orange" or "tangerine"; and there is "focal" orange, or the center of gravity, color-wise, for all the oranges of all values and intensity that people sort into the broader category we call "orange"—everything from apricot to creamsicle to sunset to burnt orange to safety orange to orange peel and so on. None of these colors have to be, or are, the same: Color specialists note that typical green, based as it is on plant foliage, is often yellower than unique green. This difference between unique and typical becomes a massive problem when you start piling colors on top of one another in a description. "Khaki," for instance, would have been described in the old

ISCC-NBS method as a "moderate medium yellowish yellow orange." The vast majority of people will not pay much attention (initially) to "moderate" or "medium": They're going to key in on "yellowish yellow orange," beginning with "yellow orange." They will understand that "yellow orange" is a combination of yellow and orange, and that "yellowish" means it's more yellow than yellow orange. They will mentally combine, or attempt to combine, their typical yellow and their typical orange into some sort of color between the two, and then tip it yellow-ward. But the ISCC-NBS system doesn't rely on *typical* colors; it relies on *unique* colors. Accurately decoding this plain-language definition requires a *Munsell Book of Color* (re-notated), the suspension of everything you know about "yellow" and "orange," and about fifteen (preferably caffeinated) minutes of your time to figure out that "khaki" is *tan,* which is a type of *brown,* not a yellow orange.

To make the ISCC-NBS method useful to as many people as possible, the method had to allow for a broader group of common colors. After another survey of all the color names in use, the original basic colors (red, orange, yellow, green, blue, purple) were expanded to include "brown," "olive," "violet," and "pink." Those core colors could combine using one "-ish" modifier—"yellowish brown" or "bluish violet"—and that gives you a color neighborhood (what the ISCC-NBS folks called a "pocket") that's easy for an average person to visualize and gets them colorimetrically closer to where the ISCC-NBS needs them to be.

There was another important discovery made by the Problem 2 committee as they began to apply their defining rubric to new sets of colors. The committee realized that their first modifier grid, with clean lines between "dark" and "medium" and "light," and "strong" and "moderate" and "weak," didn't fit the same way over each part of Munsell's color space. If the ISCC-NBS was supposed to follow common usage, then they had to reckon with the fact that, for instance, most colors called "dark teal" were, by measurement, a lot lighter than most colors called "dark brown." The ISCC-NBS committee began mapping all their collected color names to find the boundaries between the initial descriptors on each slice of Munsell's color solid. Godlove's initial tidy grid was, like Munsell's initial color sphere, skewed, squashed, stretched, compressed—over and over and over again for each color. And in doing so, they found, thanks to the rapid growth of color in the consumer world, new modifiers: "pale" for colors that were grayish and light, "brilliant" for colors that were strong and light, and so on. They collected, mapped, and refined.

Godlove explains all of this ongoing work in one of his 1948 letters to the Merriam offices. "In doing this [the ISCC-NBS team] has practically followed the Merriam-Webster's own principle of following, not leading, popular usage," Godlove assures a silent Merriam-Webster.[1] And regardless of whether they're interested or not, he's moving forward as if he were defining for them. In his copy of the NBS colorimetrist Deane Brewster Judd's 1949 progress report, which includes a new Munsell chart for one of the hue ranges (1R–4R, reds and pinks) with the new nomenclature, there are a handful of color names plotted carefully in their respective pockets in Godlove's distinctive loopy hand: "begonia," "rose d'Althaea," "carnation rose," "sea pink."

In June 1950, while the Merriam Editorial Board was welcoming Twaddell, Godlove wrote to Oakes to let him know that the newly revised ISCC-NBS system had been officially adopted by the delegates of the ISCC, and just to show Oakes how absolutely perfect the standard is, he enclosed 841 definitions of color names put together based on the new ISCC-NBS style—which he's put together in spite of some "considerable ill health (not serious, but rather annoying and time-consuming)." He explains the system thoroughly to show how usage-minded it is: ordinary color names, easily understood modifiers taken from common use, each word in the system beat to death by a decade of committee work. Each definition, Godlove reckons, will average only four words, which frees up space for perhaps thousands more color names than were entered in the *Second*. Everyone would be happy—especially Godlove. "I was chairman of the committee which instigated the color-names work twenty years ago, and national chairman of the Council when the system was finally adopted. Thus it is pretty much my brain child and I am anxious to see its universal adoption."[2]

Oakes received Godlove's letter with more than a tiny bit of consternation. While the ISCC has decided that terms like "deep" and "pale" are scientifically accurate, are they, really, in actual usage that accurate? He was dubious. Here was one good thing about the letter: It came with a great sheaf of already-defined color names, and they didn't need to pay a cent for them. (Yet.)

With criticisms of the *Second* ringing in his ears, Oakes was wary in his response. The ISCC-NBS system is much better than the one they had come up with for the *Second*, from a sheerly lexicographical perspective. But he wasn't convinced. There are thousands of colors spread across three thousand pages; are these new terms so intuitive that a reader wouldn't need to flip back and forth to some key somewhere to know that "brilliant" referred to a color that was strong and light? "What may seem like obtuse reservations result from our normal attempt to place ourselves in the position of the casual consulter of the dictionary," Oakes explains.[3]

He then did something colorimetrically smart. He asked Godlove to give them about a dozen samples of these new definitions to illustrate the various types of color entries, including a bunch of problematic edge cases: "eggshell," "off-white," and "aqua." They'd evaluate and let him know. Ten days later, Godlove wrote back and sent along almost four hundred definitions for review—

and no "aqua" or "off-white," which Oakes had specifically asked for. Oakes was exasperated, but the unasked-for samples were instructive.

First, they showed exactly what kinds of color names Godlove was planning to enter. Very few surprises there, though a handful of entries raise some questions. There are some color names that look, to the outside observer, to be variants of the same color, like "cinnamon" and "cinnamon brown," but they are defined by Godlove as two different colors. Why? Upon what authority? Some entries seem to have two definitions: one that's unmarked, and one that's marked "TC," apparently because that's the color used by the Textile Color Card Association, which tracked fashion colors. So what's the unmarked definition for? The definitions themselves were straightforward enough, but some included a "to" in there: "moderate to strong red." Does this imply that this color name is used to describe a bunch of similar-yet-distinct colors? Is the "to" marking a range, and if so, where? How is that specific?

Quibbles aside, this is a good defining paradigm to start with, but there were still some commonsense problems to hash out. Oakes centered on one of Godlove's definitions for "lilac." It is "light grayish red."[4]

Though this definition is completely accurate according to the ISCC-NBS method, it is just wrong to the Average Color Viewer. We have such strong associations between intrinsic color names like "lilac" and our generic descriptions of them that our mental color categorization of them is fixed. It doesn't matter if, colorimetrically speaking, the color that Godlove is calling "lilac" belongs in the "grayish red" pocket or not: We are accustomed to thinking of lilacs as purple, and therefore the color must be defined in reference to purple as well. It's right there in the definition for the plant, for Munsell's sake: "a plant of the genus *Syringa; especially*: a European shrub (*S. vulgaris*) that is often found as an escape in

North America and has cordate ovate leaves and large panicles of fragrant *pink-purple* flowers." The Average Color Viewer might not recognize many of the words in that definition, but they sure as shootin' know what "pink-purple" is.

This becomes a little more ridiculous when the color name includes a basic color term in it and isn't defined according to that basic color term. How can "Nile *blue*" possibly be a "light bluish *green*"? Yes, yes, language is not logic, but goddamn it, do you have to make it so flagrantly illogical? We have so tied the idea of the basic color categories like "blue" being described universally as, well, *blue* that if a language (like ancient Greek) doesn't have a one-to-one matchup for our word "blue," people will claim that the people who speak that language can't actually *see* blue.*

Oakes also notes something problematic in Godlove's formula. The ISCC-NBS standard is going to divide all colors up into about 270 of these color neighborhoods, or "pockets," and each pocket is going to get a simple formulaic definition. Given that the Average Color Viewer can see (Godlove claims) up to ten million different colors, this means that hundreds of colors will all have the same definition. "Crimson," "scarlet," and "vermilion" would all be defined as "vivid red," and "old gold" and "citrine" would be "light olive brown."

But this is a lexicographical nightmare. The *Third* was to be an unabridged dictionary, and people expect fine-tuned lexical differentiations between similar words in an unabridged dictionary. If "crimson," "scarlet," and "vermilion" are defined as "vivid red," then one of two things will happen. One, the reader will assume that all three of these words refer to the same exact color—that

* Let's put this to rest. The ancient Greeks carved up the color solid into different sections than modern English speakers have, which means that ancient Greek actually had *more* basic words for blue than English does. One of those words is the root of our word "cyan," goddamn it! I will yell about this until I am *glaukos* or *kyaneos* in the face.

they are synonyms of one another. Or two, the reader will know somehow, deep in their waters, that these are all different colors, and call you out for being an idiot.

There is a limit to what the dictionary reader will accept as not just accurate but lexicographically elegant. Accuracy means nothing if there's anything in the definition that makes the reader stumble, or, worse, stop entirely to consider whether the compiler of this dictionary was a witless dingus. A well-written definition should be as bland as dry toast: nothing to write home about, nothing you crave, but stabilizing and predictable. To balance accuracy with this sort of inoffensive smoothness is an art, and it is not one that Godlove has mastered. He did eventually send Oakes a definition for "off-white." His definition: "a color resembling white, but somewhat reddish, yellowish, greenish, bluish, or purplish; or darker or both slightly darker and chromatic."[5] Smooth it ain't.

Twaddell was out; Gove was swept in. With a change in regime came a change in philosophy: Consultants were fine, but the *Third* was to be a dictionary led by *lexicographers*.

Gove knew that given the way that the world was moving, coverage of scientific and technical terms in the *Third* needed to be as rigorous as it was in the *Second,* if not more so. This had to be the province of the outside consultants; though Gove had abandoned academia for lexicography, he understood that wasn't practical for most people. But Gove didn't want those consultant-written entries to go into the *Third* without some sort of extensive internal review by in-house editors. This was a different tack from the *Second,* which plainly states, "The definitions in these special subjects have all been entirely written, rewritten, or otherwise checked, by the Special Editors."[6] The inconsistencies of the color definitions in the *Second* were due not to Carhart's ill health and inability to check everything but to the idea that the "special editors," as the consultants were called, were always right. Gove felt that the consultants had been given too much leeway.

The issue was one of focus. Trained lexicographers knew what sort of information was essential to a dictionary entry, and what sort of information was stuff that users wouldn't need to know to understand the meaning of the word at hand. This is a departure from the *Second,* where the focus wasn't on the lexical meaning of a word in this context but on capturing the entirety of the thing being described:

> **1.** *Chem.* An element occurring free as a colorless, tasteless, odorless gas (ordinary oxygen) in the atmosphere, of which it forms about 23 per cent by weight and about 21 per cent by volume, being slightly heavier than nitrogen. Symbol, *O;* at. no. 8, at. wt., 16.000. Oxygen is the most abundant of all the elements on the earth's surface, for, in addition to its occurrence free in air, it forms, in combination, eight-ninths by weight of water and nearly one-half by weight of the rocks composing the earth's crust, being a constituent of silica, the silicates, the carbonates, the sulphates, etc. It is also a constituent of a large proportion of organic compounds. It is indispensable in respiration. Ordinary oxygen was discovered in 1774 by Priestley (who called it *dephlogisticated air*) and independently by Scheele. The pure gas may be prepared by heating potassium chlorate (best with addition of manganese dioxide), and by various other methods. The chief industrial methods are the liquefaction of air and the electrolysis of water. Sp. gr. referred to air, 1.105. Weight of a liter, 1.429 g. Molecular formula, O_2. It is used in oxyacetylene and oxyhydrogen blowpipes, in metal cutting, in liquid-oxygen explosives, in purifying illuminating gas, in medicine to aid respiration, and for a few other purposes. Oxygen can be condensed to a pale-blue liquid boiling at -183° C., and to a pale-

> blue solid melting at -218.4° C. The liquid is strongly magnetic. Ordinary oxygen is moderately active at ordinary temperatures, and its activity is greatly increased by heat. It combines with all the elements except those of the argon group and fluorine, the process being called *oxidation,* of which ordinary *combustion* is only an intense form. The element oxygen is commonly bivalent (but see OXONIUM). It is a constituent of all but a very few acids (as hydrochloric, hydrobromic, etc.), and, in general, the greater the proportion of oxygen with which an element combines, the more acidic does it become. Oxygen is also known in an allotropic, more active form, *ozone* (which see). See also OXOZONE.*

That is the entry for "oxygen" in the *Second,* written by Dr. Austin Patterson, vice president of Antioch College and former editor of *Chemical Abstracts*.† It's an entry that is literally breathtaking in its exhaustiveness, and certainly supports the hype that the *Second* was the sum of all human knowledge condensed (such as it is) into one tome. But about 90 percent of this entry is exactly the sort of space-wasting verbosity that Gove wanted eliminated—pretty much everything after the word "atmosphere" was subject to Gove's red pen.

And so all the specialty defining turned in by consultants was to be reviewed and evaluated by a handful of editors, including Oakes, Mairé Weir Kay (a no-nonsense science editor who, until the day she died, was known around the office as Miss Kay), and Hubert Kelsey (who disliked Gove immensely and didn't particularly think much of being told to babysit a bunch of overconfident and conceited consultants, either). Anything that made it into

* There is a second definition at "oxygen": "2. Chlorine used in bleaching. *Manufacturing Name.*" It's easy to overlook for obvious, obviated, bloviating reasons.

† Also, it's worth noting, a man who approached Albert Munsell on behalf of Merriam-Webster back in the early twentieth century and tried to persuade him to put his *Color Notation* into the 1909 *New International Dictionary.*

the *Third,* Gove made clear, was *their* responsibility, and not to be blamed on a bunch of by-blows who happened to have advanced degrees in this field.

Oakes knew that they had little choice at this stage in the game but to try to reinvent the color wheel, lexicographically. Here's an eager beaver who's already handed in a bunch of work; someone who was familiar with lexicography, had worked with the company for the *Second,* and had presented himself at the right time. It was a gift horse, but Oakes couldn't help but stare hard in its mouth. He eventually threw up his hands. "I do not see how we can do otherwise but to use the defs of the Color Council in the 3d Ed," he wrote in a memo to Bethel.[7] It is not exactly a full-throated endorsement.

In Living Color

Part of Gove's vision for lexical excellence involved cataloging the minutiae of how, exactly, a dictionary should be put together and under what principles. (It was, in his opinion, a better use of his time than dealing with running an editorial department.) He collected his and others' thoughts into a multivolume style guide so dense it rivals neutron stars: These volumes are called the Black Books.

The Black Books touch on literally everything required to write an unabridged dictionary, including the prewriting parts, and quantifies, qualifies, and data-drives points as diverse as finding evidence for definitions and when to use a boldfaced colon in a definition and why. The more an area needed clarification, the more memos were written about it. One such area gets an all-caps heading in the index to the Black Books: "SUPPORT FOR DEFINITIONS." Ostensibly, every word defined inside a Merriam dictionary was entered based on that word's use in edited prose—what we call citational evidence in the biz—but in the case of the *Second,* it was almost impossible to prove. Hundreds of outside consultants contributed definitions that had no citational evidence attached to them; there was, prior to Oakes's reading program, no consistent collection of citational evidence by in-house editors; the in-house editors had no way to check the specialist definitions—or each other's definitions—against the word's actual use in print. This did occasionally cause some embarrassment. "Pneumonoultramicroscopicsilicovolcanoconiosis" was added to the 1939 New Words supplement of the *Second* based on a single citation from Frank Scully's *Bedside Manna,* which described it as a medical term for a lung disease caused by the inhalation of fine silica particles. What the definer who drafted the entry did not know was that "pneumonoultramicroscopicsilicovolcanoconiosis" was not actually in use in medical texts: It was created by Everett M. Smith, the president of the National Puzzlers' League, who coined the word as the longest feasible word in the English language and introduced it at the league's 103rd meeting in 1935.[1]

In an effort to explain just how data-based the *Third* was going to be, Gove's memos outline how many citations were needed to justify entries in the *Third;* the citational criteria varied extensively depending on the type of word being entered. But Gove insisted that, yes, we needed several typewritten pages on the topic to make sure that there was no confusion on anyone's part. No more trusting editors to have looked through the citations in the catalogs while writing; all citations used in the writing of a definition needed to be literally attached to the defining slips. The memos reiterated over and over that all definitions in the *Third* were to be based on an analysis of the collected citations for that word. No more relying on the lexicographer's intuition: A dictionary of "pure language" has to be written based on evidence, not vibes.

So by the end of 1952, when Oakes wrote Godlove back, part of the training that all the newbies settling into their desks were getting from Oakes was how to read and analyze citations in order to write definitions.* But how were the in-house editors supposed to verify that what was being turned in by the consultants was, indeed, backed up by evidence?

A potential answer to one consultant's word list had dropped into Oakes's hands, courtesy of the reading program. Walter Granville, a well-known color scientist, had recently sent Merriam-Webster a publication he had worked on, to be published in conjunction with a new color standard that focused on mass-market colors. This was the *Color Harmony Manual,* published by the Container Corporation of America in 1948, which sprang out of discussions not between scientists, as the ISCC-NBS standard did, but between color psychologists and color consultants who worked in retail, primarily. Granville worked with Helen D. Taylor and Lucille Knoche, both color consultants for a number of retail outfits like Sears & Roebuck and Montgomery Ward (among others), to write a companion *Descriptive Color Names Dictionary* to be packaged with the *Color Harmony Manual.*

The *Color Harmony Manual* was, right around the time that Godlove was writing, having a moment. It wasn't just a standard like the *Munsell Book of Color,* or a method like the ISCC-NBS, but included work sheets and exercises to help teach color theory, and was therefore popular with artists, architects, designers, decorators, printers, schools, libraries, fabric manufacturers, the plastic industry, and makers of paint, paper, and wallpaper. The *Descriptive Color Names Dictionary* in particular piqued Oakes's interest: He noted in his letter to Godlove that it was used by "mail order houses," which would "provide valuable listings for our pur-

* Gove himself was far too stressed out to do the actual training; he was not the sort to, while highly pressurized, carefully sculpt a savvy lexicographer from the slab of smooth dumbness that he had just hired. No disrespect to the editors hired for the *Third:* We've all been that slab.

poses."[2] Here, at last, was a color dictionary based on actual usage. And what usage, indeed.

The 1952 Sears & Roebuck Christmas catalog: 440 pages long, with lots of four-color pages to better tempt the buyer, and carrying everything from clothes to beauty supplies to home furnishings to sporting equipment to accessories, and bane of Godlove's current lexicographical existence. The menswear section is up first, and the first page lists dress shirts, helpfully illustrated with a jolly office worker in a Santa hat. The colors are all very subdued, pastels, light. The color names: "green," "gray," "blue," "white." A few pages over, ties get a little spicier: "mint green," "maize," "chocolate brown," "scarlet red." Flip to page 6, to the casual-wear work shirts, it's more of the same: "rust," "maroon," "pearl gray," the beautifully tautological "gold color."* And most of the menswear follows this color-naming pattern: The colors are all either very common (maroon, light blue) or easily identified with something in the real world (chocolate, maize).

Womenswear, by contrast, uses names that are far more evocative. Ladies could get nylon and silk stockings in "burnished beige," "moonstone," "towntaupe" (one word), "white," and "black"; wool cardigans in "winterberry," "rico green," "amberhaze" (one word), "wine flame," or "toast." Even office wear got more whimsy on the other side of the catalog: For the men's "lt. blue," women got "powder blue," and for "green," they got "mint green." Ladies could have blouses in "soft pink" and "soft gold," which were distinct from "pink" and "gold" in that they were . . . soft.

* If Sears is to be believed, men didn't wear yellow in the 1950s unless it was in plaid. They wore only gold and maize. If I were a turd stirrer, I'd say that this meant that men couldn't see the color yellow until the 1960s.

Catalogs like these, advertisements for consumer goods, and the fashion columns in women's magazines (and a handful of the dailies) were, for many people, the main point of contact with color names. And these catalogs, advertisements, and fashion columns were meant to do one thing, and one thing only: whet the consumer's appetite for more stuff. Four-color printing technology was in its infancy, so colors seemed garish, oversaturated,* too bright. They weren't always a good representation of the color at hand, and so these advertisers used the next best thing: color names.

The more evocative the color name, the better. "Josephine," for instance, is a darker blush color that is not named for any particular Josephine, but the name itself is feminine, French, *refined*. "Rose blush" is another name for Josephine, and it is certainly more descriptive, but real talk: A dress in "rose blush" sounds like something that your grandma or the mother of the bride might wear, whereas a dress in "Josephine" is romantic, sweet, *French*. You might not pay $100 for a grandma dress, but you'll think $100 is a steal for a *French*(-sounding) dress. Remember the types of color names from earlier: basic, intrinsic, associative, and fanciful? Color psychologists and market researchers have studied how people react to color names in retail settings, and time and time again these studies show that people are more likely to purchase an item with an associative or fanciful color name over an item with a basic color name, regardless of what those actual colors are. And, more to the point, they're also willing to pay *more* for the product with a fancy color name. One study even calls this the "Fancy-Name Effect."[3]

Fashion being a luxury good, it could afford some really frilly names. One of the fashion forecasts for 1952 (fall, for man-made

* Excuse me: too chromatic, in that they appeared to have a higher chroma once printed than the target color. Let us, even when discussing fashion, be as precise as possible.

fibers and silks) includes names like "Kingfisher," "Goldflake," "Sugarpine," "Copperglaze," "Wood Iris," "Cheeryglo," and "Azurtone."[4] These names might bring to mind a color—"Goldflake" is probably in the yellow family, and one assumes "Copperglaze" might be, you know, coppery—but the names aren't meant to dump a reader into the appropriate Munsell color pocket but to get them to open their pockets. Fancy naming was taken to some extremes very early. One 1926 color forecast from Cheney Color Service featured colors named by the women in the office—colors like "Ambroon," "Bluridge," and "Flambic," names that sounded modern but were so detached from any particular color that they could be applied to future color trends, used and reused as needed.[5] Journalists who were reporting on the couture fashion debuts rarely stopped to ask the designer what they'd like that color to be named in print, but went ahead and wrote from their own memories and using their own stylistic flourishes. Even when journalists had a piece of fashion in front of them and accurately reported the manufacturer's color names, they often introduced other problems in describing the colors. Lois Long, the fashion writer for *The New Yorker* and the woman who the editor William Shawn famously said invented fashion criticism, described the 1956 fur color "Autumn Haze" as "the warm tan of a beagle puppy," which is less a color name than a color feeling, and is a slightly tweaked comparison considering the object under description is a fur.[6]

This idea that whimsical or evocative color names would better sell a product wasn't just restricted to textiles. In 1950, Helena Rubinstein debuted its Sensational Silken Lipstick in twelve (very similar) reds, from "Crackerjack" to "Command Performance"—the latter also being the name of a perfume by Helena Rubinstein, which, for the record, was not a dark red. Ads for the new lipsticks were placed in all the best magazines; it was clear, even with the best four-color printing that fashion money could buy, that the twelve reds were all very, very close together, colorimetrically speaking. With such a limited palette of colors on offer, the name was the only way to nudge a buyer toward Copper Leaf (a saturated, warm red) as opposed to Crackerjack (a nearly identical saturated, warm red).[7]

Ironically, the one person who can likely be blamed for the proliferation of these froufrou names is the same person who acted as the initial clearinghouse for all the color problems for the ISCC. Margaret Hayden Rorke was an early member of the ISCC; she was not a scientist but a fashion forecaster. She was hired by the Textile Color Card Association in the 1920s to help them come up with fashion cards for American retailers; by the 1950s, she had built a subscription-based color-reporting empire using everything from marketing savvy to embedded spies in Paris to report on the most au courant color trends in Europe. Their fifteen hundred subscribers, most of whom were manufacturers of fashion or accessories, ate it up. Here was a detailed report about which direction Paris's colors might be heading, with enough lead time for most retailers to start production on whatever fabrics, leathers, felts, and plastics they'd need to have on hand when these fashions finally crossed the Atlantic, as well as analysis on how those colors would be received in the United States. The TCCA color cards served as the baseline for anyone producing a color consumer good, and by the 1950s, if you were making anything intended to be bought by a human being, you were likely producing a color consumer good. Rorke invented the color forecasting industry, which became, very quickly, both wildly profitable and very crowded.

Because Rorke was a member of the ISCC, there was no way for the scientists among them to ignore the psychological effects of color names. Some of them, in fact, tried to approach the psychology of color from a scientific point of view. After all, science had already broken down our experience of color into three discrete phases: the physical refraction, the biological event, and the psychological result. We could accurately and minutely measure color as a physical refraction; we could now identify the specific biological structures within the eye responsible for color vision (and problems with it); why should the psychological sensation that results in "seeing" a color be any different?

Enter Faber Birren. As a student at the University of Chicago in the 1920s, and later working for a book printer, he studied color theory and was fascinated by the effects that different colors had on people. In the 1930s, he started his own color consultancy, where his focus was primarily on using color in manufacturing and industry.

He was in the right place at the right time. Manufacturing boomed as soon as World War II began—so many munitions and uniforms and tanks and planes to build. But the folks working in these new factories were for the most part unskilled, and accident rates skyrocketed. Birren, as someone who was interested in the function of color on the psyche, was called in by the army and then the navy to consult. He suggested painting walls and machinery colors that reduced eyestrain—he recommended a light green*—and advised that any dangerous part of a machine be painted a bright color, like red or orange. The contrast would stand out, thereby keeping tired or inexperienced machine operators from jamming their fingers into gears and such. This approach was later called functional color, and it worked: When the army and navy put his suggestions regarding safety colors into force, their accident rates dropped precipitously.

* The walls on the editorial floor at Merriam-Webster were painted this same light green in the late 1940s. If ever there was a workplace that induced eyestrain, the dictionary company was it.

This might have seemed like magic to manufacturers, but it was a practical application of much of the color theory that Birren learned as a student and later a printer. He credits, over and over again, throughout his life, the groundbreaking work of the nineteenth-century dyer and chemist Michel-Eugène Chevreul, whose work at the Gobelin dye works in France put him right in the center of the Venn diagram of art, industry, science, and sales. Chevreul wrote about the experience of color and color perception; he's best known for his seventy-two-point hue circle and the concept of simultaneous contrast, which fascinated Birren to no end. In a nutshell, Chevreul said that our perception of a single color depends on and is changed by the colors we put around that color. This was something that artists knew and could manipulate to good effect: Michelangelo used it when painting the Sistine Chapel ceiling, knowing intuitively that the only way that his figures would be discernible from the floor of the chapel was by putting their red- and yellow-toned bodies against blue or green fields, or by wrapping them in cloaks and clothes of highly pigmented yellows or golds against dark greens. But Chevreul was the first to put this artistic sense of color into words. His *Principles of Harmony and Contrast of Colors* was hugely influential on artists like Delacroix, Van Gogh, Seurat, and the later Bauhaus painters like Johannes Itten and Josef Albers (whose 1963 *Interaction of Color* supplies some of the simultaneous contrast "optical illusions" that you end up seeing all over the internet now). Birren wasn't doing anything new with color and simultaneous contrast; he was just doing it on levers and cogs in the factory instead of canvas in the studio.

Birren was the go-to guy from that point forward in functional color, color psychology, and even color-adjacent sales.* His clien-

* One grocery chain in the 1950s asked Birren to come in and help them sell meat. The meat was fresh, but it looked washed out or pale under the bright lights of the

tele included the military, hospitals, schools, DuPont, Monsanto, General Electric, and the Walt Disney Company. He went on to write hundreds of articles on color and color psychology, in addition to about two dozen books, and his views heavily influenced the way that commercial colors were named and talked about.

Birren believed that the effects that color had on people were not merely physical but metaphysical, psychological, cultural, and even (regrettably, cringingly) racial. A person's response to color was so holistic and complex, Birren thought, that if you're going to market a color, your marketing had to take into account that complexity. In 1945, he released a book of his functional color studies aptly called *Selling with Color:* It was part philosophical treatise and part sales manual, and it was, for color forecasters, a hit. His approach is both scientific and practical. He manages to synthesize the findings of color psychologists and color researchers with practical sales advice: People prefer more familiar colors over "new" colors, so from a merchandising standpoint it's better to offer lots of new products or designs in a limited number of colors, rather than one single design or product in a panoply of colors; color trends in mass markets change slower than in high-fashion markets, so it doesn't do you any good to follow Parisian fashion if you're a manufacturer of, say, Bakelite radios. There's a lengthy appendix that offers actual sales figures from a variety of industries about which colors sell the best, and another that lists a bibliography for those interested in color psychology. He does stray into some sweeping generalizations, of course:

---

store. Birren suggested they paint the backdrop of their cases green; green and red were opponent hues, and so the green would make the red of the steaks pop and look fresher. You still see this in operation today. If you've ever wondered why meat cases in grocery stores sometimes use bright green plastic "grass" sheets to divide cuts, you can thank Faber Birren.

> Athletes are said to prefer red, intellectuals blue, egotists yellow, while the convivial favor orange. Through national and religious traditions, some peoples—Hindus, Chinese—look upon yellow as a sacred and happy color. Mohammedans love green, the hue of Allah. The Irishman is prejudiced against orange, and many Scottish people dislike green.[8]

Selling with Color emphasizes over and over that people's responses to color are conditioned and primal. People can't talk about color well—he notes that studies show people can name only about thirty to forty colors, though he doesn't give any citations for that figure—but they sure can *feel* color. "Form is something for the brain to appreciate, and the brain may be educated," he writes. "Color, however, is more like religion. It is in the blood, an essential part of the psychic makeup of an individual."*

Birren's theories on what certain colors said to people appealed to color forecasting companies, and one in particular: Pantone. Pantone started out in 1963 by providing color standards for graphic designers, then moved into the world of color consulting, helping choose corporate colors for clients like Coca-Cola (Pantone 185), GAP (Pantone 492), and John Deere (Pantone 364 C and 109 C). In the 1980s, Larry Herbert, the owner of Pantone, hired Leatrice Eiseman, a color psychology expert, to move the company into the fashion color forecasting world. Larry's daughter Lisa and his son Richard assisted in setting up the initial palettes for their color forecasting gig, collecting samples of colors used in fashion, heading

* Birren takes the whole "color as religion" thing a little further in his 1950 book *Color Psychology and Color Therapy: A Factual Study of the Influence of Color on Human Life*. Chapter titles are things like "The Inspired Mystics," "The Bewildered Philosophers," "The Amulet Wearers," "The Aural Healers," and "The Eager Chromopaths," and he solemnly and earnestly writes about mystical healing rites, alchemy, the astral plane, auras, and crystals as they pertain to color. It reads like Godlove on acid.

up to the attic, and literally taping them to a chart of a color space to see where the colors clustered. The 13,000 colors were narrowed down to a more manageable 1,001 after a few iterations, and then Richard organized those 1,001 colors according to the ISCC-NBS model.

Pantone had avoided using color names since the beginning, because color names were imprecise—something that Birren notes repeatedly in his writing. But part of color forecasting involves grabbing a consumer right by the feels, and a palette consisting of colors named 187, 306 C, and 145 just didn't have the requisite oomph. Lisa, Richard, and later Eiseman needed to give their palette colors names. They started with the 267 central pockets of the ISCC-NBS, aiming at giving them color names that were poetic, or that conjured specific color memories: Pansy Violet, Lavender, Zuni Brown. These, more than a set of numbers, conjured a feeling, and that's what you want a consumer to buy. "It's important to note," writes Eiseman, "that as much as 95% of consumers' decision-making is dictated by the subconscious."[9]

But associations and feelings change, which means, then, that the names we give colors must change as well. Eiseman explains:

> These new names are created for the press and buyers. . . . When you change the name of a color, you evoke a completely new set of emotions and sensations. Forget your basic orange and green. Ignore purple. Disregard brown and wine. Now you can select from pumpkin, parsley, eggplant, chocolate or claret.[10]

That's cheating! the consumer cries, but this sort of rebranding has been going on in English for centuries. In the late fifteenth century, there was a brown whose name was taken from a very high-quality woolen cloth. The color's name was "puke." It appears quite a bit in printed prose until the mid-seventeenth century, and then its use begins to decline. This coincides with the sudden rise of the verb "puke," meaning "to regurgitate the contents of one's stomach," and the matching derivative noun "puke"—both of which are etymologically unrelated to the color name. It doesn't matter how lovely the color "puke" actually is: The rise of the new "puke" effectively stopped the color name in its tracks. And thanks to the new "puke," whenever modern readers run across "puke color" or "puke stockings" in historical texts, they assume that "puke" must be referring to the color of bile, not the original dark, red-toned brown that was "puke" before "puke."* Lexicographers know this because when "puke" now shows up in sentences about colors, it's paired with "yellow" or "green." This association is so strong that it's probably one of the things that makes people think that "puce" is a green: "Puce" sure looks and sounds like the word "puke," and we know that "puke" is green. ("Puce" is not green; it's a reddish brown. I don't make the rules.)

Even in markets where developing new colorants and pigments took years, the juggernaut of sales still rumbled forward. Manufacturers could not reliably come up with every single color in existence for every single application, but a name that evokes a feeling is also imprecise enough to allow technology to catch up. In 1951, you could get a Mercury Monterey sedan in fourteen colors, four of them greens (Sheffield, Coventry, Brewster, and Yosemite). In 1952, Mercury's Coventry Green was renamed Coventry Gray Green, which was renamed Pinehurst Green in 1953. Was this a con, tricking potential buyers into thinking they were getting the

* Shakespeare famously used the phrase "puke stockings" in one of his insult pile-ons in *Henry IV, Part 1* ("Wilt thou rob this leathern jerkin, crystal-button, not-pated, agate-ring, puke-stocking, caddis-garter, smooth-tongue, Spanish-pouch?"), and misinterpretations of the phrase abound. Students, please note that the "puke stocking" here does not refer to a stocking of a particular color, as half of your textbooks will tell you, but to stockings made of the fine-grade woolen cloth called puke. Show your teachers for extra credit!

hot new color of the season, though it was three seasons old? Absolutely not, Mercury executives might say. It's a color so nice, we named it thrice.[11]

There's another thing to contend with when it comes to the longevity of commercial color names. Like all words, color names arise from a specific context. And outside that context, they can be dangerous.

In the 1860s, thanks in part to an enterprising and rabidly self-promoting Brit by the name of Charles Frederick Worth, Parisian fashion had become the ultimate luxury item, sought after by both cultured (and coutured) European royalty and the well-dressed socialite alike. By the late nineteenth century, the fascination with Parisian fashion had spread to the States. American periodicals breathlessly reported on all the latest colors and styles to be worn by the haute coterie in Paris, and American manufacturers, always a season or two behind, scrambled to keep up with consumer demands for everything de la mode.

One of the best ways to whet the appetite for the glamour *de Paris* was to transport the reader directly to the *couturiers* of the Rue de la Maison with liberal application of *couleur locale* and *la langue française*. Not sure how to get the jet-setters of New York interested in a dusty-pink bonnet that their grandmothers wouldn't be caught dead in? Say that the color is *vieux rose* and it's no longer déclassé. "Mint" sounds bland, provincial, snoozily *American,* but *pomme verte* is a color that sighs across the lips like a lazy summer breeze off the Seine. French words have traditionally been associated with elegance, education, Continental manners and mores—and price points. A judicious application of just enough French could elevate a blah color to the belle of the ball.

And so it was with a brown that in French was called *tête-de-*

nègre. It first appeared in a *Harper's Bazaar* fashion report from 1876, where it was glossed as "a red maroon,"[12] and like all French color terms it was sprinkled among the fashion reports for the next thirty-five years, where it was also described as "the darkest of browns" and "a beautiful brown." But there's only so much French the average housewife in Tennessee or Minnesota is going to retain; by 1910, *tête-de-nègre* was being translated in American newspapers as "nigger brown."

If your stomach curdled when you read that, then you are experiencing, in a very visceral way, the social and cultural limits that some color names slam up against. The color became fashionable, so the anglicized color name appeared everywhere. Newspaper ads blared that they had it; one ad from 1914 declared that it was "the most popular color in Dress Goods today."[13] The n-word, as we now call it, was not just a common epithet in conversation, but also appeared in the news regularly. The color name was, for the most part, unremarked upon, though its use was seen as distasteful and offensive by at least some. A 1913 column in *The New York Age*, an African American newspaper, called for Black women to boycott stores that carried the color:

> The various department stores in New York and Brooklyn are insulting hundreds of their patrons by extensively advertising this new species of brown. These stores are playing a very prominent part in disparaging a race which pours into their coffers thousands of dollars annually.[14]

The writer then went on to note rather crisply that they hadn't been able to discern whether this hot new brown was "of the 'ginger-cake,' 'seal-skin,' 'chocolate,' 'molasses,' 'near-ebony,' or 'light brown' variety." The implication is clear: Why use such a hateful word when there are plenty of other options out there?

Nonetheless, others dug in. A 1914 fashion report takes a hard left turn right at the outset by literally shouting the color name in all caps and defying people to use a less inflammatory name: "It's a color and its name is properly set down."[15] The name is repeated several times in the opening, though the rest of the article is written in a decidedly less strident, more straightforward style—the typical fashion report from overseas. Clearly the writer wanted to make an unyielding point right at the outset: The color was so popular, so ingrained in the fabric of everyday life in America, all attempts to change its name were futile.

And it seemed as if the anonymous writer-crank was right: Use of the word peaked in 1914, when fashion observers in Gay Paris noted that the color was *très chic,* and it was used regularly in millinery and dress-goods ads for several decades afterward. Paradoxically, one of the worst racial slurs in English became a byword for luxury and taste when it appeared in the fashion reports. It was common enough that its potential entry into *Webster's Second* was considered in-house, with the editors eventually deciding, with some small and mostly fastidious flinches of distaste, to keep it out of a Merriam product.

But as lexicographers like to say, and as Gove himself claimed, dictionaries are supposed to tell the truth about words—accurately and unflinchingly. Given its history, its popularity, and even its import in the broader conversation about race and racism in America, then: Is this one of the commercial color names that should be covered in the *Third*?

Godlove, for what it's worth, sent no slip in for it.

Which brings us back to the lexicographer, the scientist, and the problem of the Sears catalog. Merriam-Webster wanted this dictionary to be modern and accessible, while still being the only unabridged dictionary anyone would ever need to buy. Gove's diktat was that all drafted definitions had to be backed by evidence of actual current use (or significant historical use); fortunately, more color consumer goods were being marketed and sold in the first half of the twentieth century than had been done ever before, which meant that there was lots of evidence of color names in print. And what better authority for these names than department stores themselves, those swank purveyors of everything needed for modern life? All that would be needed here is to survey all the catalogs possible, collate that data, and determine which color names were most consistent and most common. Those, clearly, deserved a definition.

But the scientist knows that the department store is often one of the last links in the great color-name chain, and that many times the color names that make it into their catalog copy are part marketing zhuzh, part business-office caution. Make it snappy, make it sing, make it sound exotic! Poof: That red blouse is now "Oriental poppy" for spring and "Turkey red" for fall—and the colorimetrist puts his head in his hands, because according to the *scientific standard* those are two different colors, and that blouse is neither Oriental poppy nor Turkey red by the ISCC-NBS standards. And what about the color names from those fields and industries *not* represented by Montgomery Ward? Should all the proprietary color names that DuPont and Ditzler created for cars be included, though they change every year? What is the scientist supposed to do when faced with the ever-changing nevergreens that are commercial color names? And why can't they all be evenly spaced out across the color pockets? Godlove grumped to Granville, "I am aware that dusty pink and dark cardinal have been well-known and important color names; but I was amazed to find 17 different 'aquas' (and incidentally, no aquamarine)."[16] This hodgepodge of imprecise colors and names was anathema to Godlove and to the precision of the ISCC-NBS standard. But this is the hodgepodge that Merriam wanted: The average dictionary user should look up "aqua" in their dictionary and find a definition that covers all seventeen different aquas in actual retail use.

Hence the knife's edge that Oakes had to walk. On the one hand, he truly had no choice but to trust Godlove when he said that the ISCC-NBS standard was going to be *the* color standard to which all artists, scientists, and industry types would subscribe, in part because it was so late in the game that there was no way he would be able to find a new color consultant in enough time to make the endeavor worthwhile. Gove had just shared with the senior editors that, per the board of directors, all the defining for the *Third* needed to be done by the end of 1955. That was just over two years; Oakes had been corresponding with Godlove for four years and had exactly zero finished entries to show for it. But on the other hand, he also wanted to make sure that the actual color names that the layperson ran into on a regular basis—those found in catalogs, primarily—would be sufficiently covered by Godlove. If Gove was hell-bent on demanding citational evidence for every entry in the *Third,* then Oakes had to make sure that colors with a significant number of citations in the files were entered. But he had no time, no staff, and no patience to chase down citations for each one of the three thousand color names that Godlove was proposing to enter (though this, according to the Black Books, is exactly what an editor on the *Third* was to do when a consultant turned in a word without citational support).

Oakes sent a letter along to Godlove that proposed—without Gove's approval—something like a mini Maerz and Paul. What about a list of colors keyed to color charts? That seems like the best way to go after all; Oakes's reservations about the ISCC-NBS description system seem to be compounded by the idea that he'd not only have to go spelunking in the citational files, and probably well beyond, to justify each and every color definition in the *Third* but also need to justify these specialized meanings of "bright," "dark," "pale," "dull." It's a risky gambit, sure to deeply upset either his boss or his consultant. Oakes finishes, "We appreciate both the technical and the prestige value of your collaboration, as well as your liberality, provided your health does not prove a deterrent."[17]

It was as honest and as straightforward a letter as Oakes could have written. A week later, he got an equally honest and straightforward response back. "It may be as well to tell you frankly the state of my health," Godlove begins his November 23 letter. "Apparently I am recovering from a 'nervous breakdown.' "[18]

This is not the sort of thing one generally reveals to a person you're trying to impress, and it's certainly not the sort of thing shared between men who were merely professional acquaintances during the early 1950s. Godlove doesn't sugarcoat the situation at all: a number of deaths or near deaths in his and his second wife's family; money worries attendant to those deaths or near deaths; five series of layoffs at General Aniline & Film, the last of which cut eighty scientists from the Central Research Laboratory, though he was spared. He was, however, sixty and in a very specialized field: What would he do if he got laid off? "This did not help my nerves any," he noted dryly. Oakes, who was also exactly sixty and worked through both recurring and chronic illness while employed in a very specialized field, understood perfectly.

Some candor, a bit of humor: It's teed up an opportunity to gently and gracefully bow out of the color defining. They are, of course, behind; Godlove is, of course, overworked; Oakes is, of course, in the full laser beam of Gove's rapacious budget- and timekeeping. But Godlove doesn't take the out. He would worry about it once he got home from the hospital, but assured Oakes, "There is not a shadow of a doubt in our minds that the ISCC-NBS system, or something close to it, is pretty close to the way the educated layman thinks and talks about color, the system being nevertheless scientifically sound and precise."[19] But he capitulated, if not eagerly. Godlove would, for the sake of Oakes and the regrettable need to have even the color names from the Montgomery Ward catalog covered, include the names from Granville's *Descriptive Color Names Dictionary* as well as the TCCA's ninth and latest edition of their color standard, which covers more lasting color names in fashion and not the ephemeral wisps of Ambroon and Flambic. This may add more color definitions to the *Third* than expected, of course, but unabridged is unabridged.

Godlove ends with a postscript after Gove's own heart: "Since writing the above I have felt much better, and have begun some actual work on the definitions at the hospital. The doctors thought it might be good for me."[20]

Red Hot

Publishing is a machine, even if it's less precision timepiece and more Rube Goldberg. The 1955 defining deadline hanging over the editorial department like the sword of Damocles was backward engineered by Gove in a detailed memo to the board of directors in November 1952. The publication of the *Third* in 1959 (on July 1, he estimates) requires that all the copy be sent to the printer by January 1, 1957; this requires that all defining be done by the end of 1955. The *Third* will contain 400,000 entries and be thirty-four hundred pages long. It is, once more, November 1952: Gove lists sixty-nine people currently working on the editorial staff, though only thirty-eight of those sixty-nine workers are able to immediately undertake defining or the oversight of the consultants' defining. This is not just a Herculean task but a Sisyphean one. An in-house memo noted that it took Merriam 588,000 hours of editorial work to write the *Second:* "Interpreted in another way, at the rate of 8 hours a day, 40 hours a week, 2,000 hours to the year, it would take one man 294 years to finish the dictionary."[1] And that was for a dictionary that was not revised from the ground up, as the "pure dictionary" *Third* would be.

Not to worry, Gove assures the board. By 1955, about 290,000 definitions will have been done; the remaining 110,000 entries that were going to fit into the *Third* were all general vocabulary, to be handled in the office. "A schedule has been drawn up whereby beginning in January 1953 a group of editors will define and revise these 100,000 terms at the combined rate of 900 per week. . . . Every editor understands that the work allotted to him is to be worked on for completion by January 1956," he writes.[2]

Allow me to indulge in a little math of my own. Assume that all thirty-eight trained editorial staff work on general defining beginning in January 1953. The schedule requires that all thirty-eight editorial staff produce an average of twenty-four definitions a week. This is certainly possible, even leisurely assuming a few things. Assume that there is no oversight work that the more senior editors will have to do for consultants, and no work on other dictionaries that are also being produced during this time frame. Assume no vacations or sick leave; assume there's no time lost to training anyone new and overseeing their work. Assume no national crises or major production problem. Assume all definers only define and don't also proofread, copyedit, or take smoke breaks in the men's room. Assume all work is on the *Third* and the *Third* is all they work on.

There is a quaint maxim, however, about assuming and what it makes of you and me.

This pace also ignores one very important thing, and that is that not every entry is created equal, and so not every entry will be defined at the same pace. Imagine that the English language is a bog-standard bell curve. That bottom 20 percent would be entries that take hardly any time at all to define. Most abbreviations fall into this category—half the time, the expansion of the abbreviation is included in at least a handful of the collected citations—as do obsolete and rare words, which are easy because there's literally no new information to use in revising the entry. If no one has used "sdeignful" since 1748, it's a pretty safe bet that the meaning is still "disdainful." A competent, fully caffeinated lexicographer with nothing else on their editorial docket might whip through a couple dozen abbreviations or obsolete words a day. On the other end of the bell curve are the 20 percent of entries that take a ridiculous amount of time to define. These include the words that are so small we hardly think of them as words—"but," "as," "for," "a"—along with a handful of very common verbs, like "go" and "do" and "make." These entries can take months of consistent editorial work. This group also includes some words whose actual use in print is a little flabby, lexicographically speaking: You can't quite tell if the writer actually knows what the word means. "Paradigm," "god," and "esoteric" are good examples of this sort of entry; the citation files are full of sentences like "ushering in a new paradigm of growth" or "worshipping god" or "one of the more esoteric scholars in the field," featuring words whose meanings are as amorphous as a John Carpenter monster.

So much for the outliers: The remaining 60 percent of the entries are everything else, some of which tend more toward the "twenty a day" end of the range and some of which tend more toward the "I long to escape this dismal plane of existence and become a being of pure light now, thanks" end of the range. You can't guess where certain words will fall on that middle part of the curve, either. Maybe I'm having a great week and the entries for the noun *and* adjective "general" are done in a day, maybe two; or maybe my dog has died, my wife has left me, I have several pieces of really in-depth correspondence to answer, including three complaints about the entry for "sea squirt" and one query asking for a comprehensive history of the alphabet, and the batch of words I'm defining includes "depressing," "destruction," and "dismal."*

* At least "dismal" ends on a slightly up—if inane—note with "dismal Jimmy," a British slang term for a man who is known for his depressing predictions or his

Now stop assuming, and up that toll to fifty definitions per available editor per week, and Gove's tidy and untested projections start to fray and catch fire.

What Gove does not explicitly say to the board: The 290,000 words that he said were in the bag would be defined outside the Merriam offices, as was done for the *Second*. What Gove does not explicitly say to the editors, especially Driscoll, Kelsey, Kay, and Oakes: Those outside definitions need to be buttoned up pronto. Time to ride herd on your consultants.

In May 1953, Oakes reached back out to Godlove. He's solicitous about his health, so rather than explaining that he needs to kick it into high gear, he gently encourages him by noting that they will need a completed section, and especially the letter *C*, "before long." Oakes is hoping for a progress report on the defining paradigm and procedures that Godlove will use for the *Third*, but

> as always when we write to you about this project, we find ourselves revolving questions of both policy and method and querying you on point after point as if we had forgotten that you are attempting to complete your recovery from a period of trial.[3]

Not exactly what Gove would have bark-dictated at a consultant, but Oakes had his own reasons for being solicitous of Godlove's health. He wouldn't read Godlove's response for some weeks: He ended up in the hospital as well. But of course: It was practically an employment milestone at Merriam-Webster that every senior editor, at some point, achieved. Overwork is a wrecker, and Oakes had more complicating health factors to manage. But Gove's schedule was immovable.

---

generally pessimistic frame of mind. It's always a welcome break to go from reading about "dismal stock prices" or "dismal weather" or "dismal job prospects" to reading about "naysayers and dismal Jimmys."

While Oakes was out recovering, Godlove responded that he was not back up to speed entirely; he writes that his "health is much improved; but I still have the memory of two 'breakdowns' totaling ten weeks in hospital alone, primarily from exhaustion due to self-imposed overwork."[4] But he is ready to start the dictionary work more fully, and finally gives Oakes what he's looking for. After nearly eight years of back-and-forth, three years of monthly progress reports on the ISCC-NBS, and three editors in chief, at last: spread out over three incredibly dense, single-spaced letters sent in June and July 1953, a defining paradigm for the color terms. It's not perfect, he warns. It's not going to be without some gnashing of teeth, or some unhappy compromises. But it will be a good starting point.

The ISCC-NBS report was nearly complete at this point, and it would be the guiding light for these definitions. Godlove and his wife (acting as assistant) will index all of the color names that the ISCC-NBS group has gathered from their thirteen authoritative standards; that will be the starting point for his word list. Because Merriam is so very concerned with making sure that each of the colors will be differentiated, he'll begin with the ISCC-NBS rubric—"a vivid red," "a moderate yellowish green"—but then will compare that color to at least two other commonly known colors in that same pocket, noting differences in hue, lightness, and saturation. This took the best of the *Second*'s defining rubric—giving you extra information about the characteristics of the color you're looking up—but refined it by orienting the reader inside a color space by giving them another common color as a reference point. The comparisons would be made scientifically, using Munsell notations, but for the sake of comprehensibility, some allowances would be made—all using common words, using their most commonly understood meanings (whether those meanings were scientifically accurate or not). He'll use "slightly" as a modifier if the color being described was only, well, slightly bluer / redder / greener / yellower than a comparison color. In the rare cases where one color name, like "coral" or "aqua," was used to describe a bunch of colors, he'd find the coral or aqua that was smack in the middle of all the collected corals and aquas and call that "average." When the color name didn't refer to a tidy dot in the color pocket but a little smear, he'd note that with definitions that began "a variable color averaging . . ." And when a color name is applied to a tightly clustered group of colors that cross one of the boundaries between these 267 color pockets, he'd note the range with a "to": "a light brown to a moderate yellowish-brown."

Now for the preemptive naysaying: There are lots of potential problems with all this, Godlove notes. Those supplementary distinctions will have to be measured for each color he's defining. It does no good to eyeball the difference between eggshell and cream or burnt almond and tan: That's not scientifically rigorous. Some of the names in Granville's *Descriptive Dictionary of Color Names,* like "bright aqua" and "dusty aqua," are for colors that Godlove doesn't think exist in the real world. He doesn't explain why or give justification; he's the expert, after all. And, of course, Godlove would need to get rid of all the obsolete and rare color names that might show up in these sources, and that will take time. If this is to be an up-to-date, scientifically rigorous dictionary, then there's no sense in keeping the names of colors no longer used.

This work will—must—be done in his downtime from his paying job at the CRL, but he anticipates this won't be a problem. He also asks that his wife be paid $1 per hour for her indexing time.

But he's excited nonetheless. Later, he assures Oakes even if they ended up not having the space for the full definitions and so were stuck with using the base ISCC-NBS definitions, it will still be a vast improvement over the slippery, subjective definitions of years past. He encloses two handwritten charts in a letter to show just how precise the ISCC-NBS system could be using only plain language. Take, for instance, some of the greens mapped out in the ISCC-NBS system:

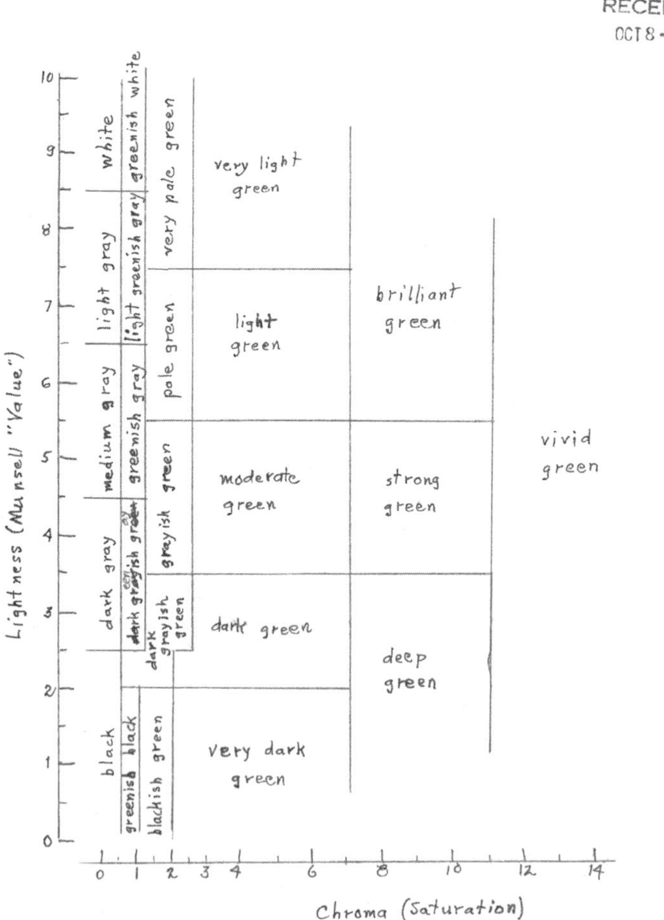

Godlove's hand-drawn map of the modifier set for the Munsell hues 3G through 9G, otherwise known as green. Courtesy of the Merriam-Webster in-house archives.

You can see, just by looking at the words, how evenly distributed this wedge of green is across both the value and the chroma scales. Now compare it with orange:

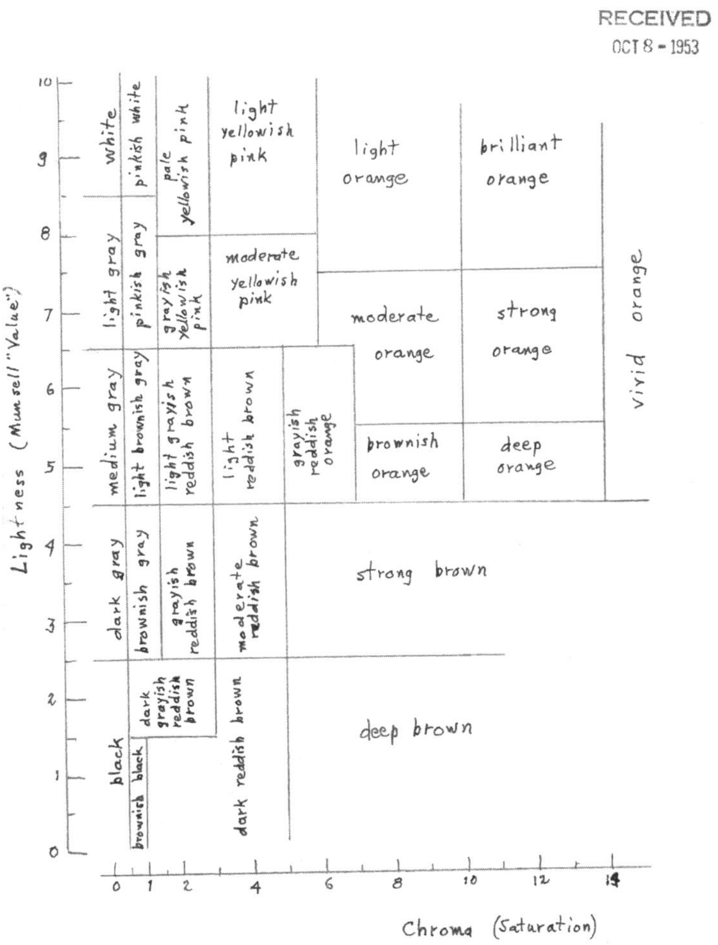

Godlove's hand-drawn map of the modifier set for the Munsell hues 2YR through 3YR, or what slobs like us call orange. Courtesy of the Merriam-Webster in-house archives.

Look at all those colors—pinkish white and light reddish brown and yellowish pink—and you may wonder if Godlove messed up and sent the wrong chart. No: This is the ISCC-NBS chart for one of the orange pockets. The colors we describe with the term "orange" are all concentrated in the top right quadrant of this chart. The further left and higher up we go, the more this slice of the color space varies: from orange to yellow to white. Head south, though, and you are immediately dumped into the depths of browns. But colorimetrically? Orange. Compare the charts themselves and you see that none of the dividing lines for these pockets of color fall in the same place for every hue. This is the sort of delightful multicolor system you get when you merge these two seemingly opposite forces: the democratic chaos of language and the curated precision of science.[5]

The stage was set. Indexing of color names began in earnest in June 1953 even as Godlove hemmed and hawed about Granville and "average"; the first slips for definitions showed up in August. Oakes had still not been able to get Gove to approve this general approach, but he had no choice. He felt that the time he was going to be able to devote to Godlove was running out.

He had no idea how right he was.

Every industry has its euphemisms, and lexicography is no exception. There's a phrase that pops up again and again in Merriam-Webster correspondence that is meant to gently explain to a writer asking for a very long explanation on a minor point of pedantry that, no, we're not gonna. We give the answer we have time for and then say that we can't continue the correspondence any further due to "the press of editorial duties." The metaphor is apt: You, the editor, are laid upon the cold metal slab while the inexorable iron plate of deadlines slowly descends, like something out of an Edgar Allan Poe story, and crushes you by minute degrees into a pulp of stress and eyestrain.

The press of editorial duties was a-cranking, which meant that everyone was getting crankier. More editors and editorial assistants were coming in and needed training. Consultants were being added exponentially, which meant that the specialty definitions that editors had to review and revise were also pouring in like water into a sinking ship. And watching the plate descend upon him was Oakes.

Oakes had begun to suspect after his hospitalization in May that his days of doing this type of editorial work were numbered. The National Institute of Neurological Disorders and Stroke, in its information on post-polio syndrome, notes it is degenerative and is "rarely life-threatening, but the symptoms, which include muscle weakness, fatigue, atrophy, and scoliosis, can significantly interfere with a person's ability to function independently."[6] Oakes was a typical patient: He wore a special brace to help counteract the worsening of his scoliosis, and he requested more time off in an effort to manage his increasing pain. Yet he was constantly worried that he was slowing down the editorial schedule and left small, apologetic asides to this effect in note after note. Gove was constitutionally incapable of believing that the editorial work at Merriam was anything but rejuvenating, but Oakes knew that the day would come—likely very suddenly and very soon—when he wouldn't be able to keep up this pace forever.

Managing Godlove wasn't even his biggest worry. He noticed, as he went through his own in-house defining work, that other people—including outside consultants—had ignored that color was being defined by an outside specialist, and so every Dr. Thomas, Richard, and Harold with color vision and a spare few minutes decided to do Merriam a solid and help "fix" the color definitions. Rather than scurry through the defining slips right then and there, and violently tear out any color definitions that weren't Godlove's, Oakes decided it would be best to write a defining memo and just emphasize to any of the folks handling the copy that if a color slip wasn't stamped "GODLOVE" and "OAKES," it was to be ignored. Oakes's two-page memo begins, "Godlove's psychophysical defs of color and their references had better be regarded as sacrosanct."[7] That was not a request; it was a threat.

Gove, too, grew more pinched. In early 1953, there was a change in the guard: Gordon Gallan, previously the chief marketing man, became the new president of Merriam-Webster. Gove saw an opportunity to get out from under the weight of managing people. As soon as Gallan was promoted, Gove proposed that the board of directors counteract the widening division between corporate and editorial and address the increasing dissatisfaction among the editorial staff by making Gove a vice president. Nothing would show the editors that the board took them and their concerns seriously more than including a member of the editorial department on the executive team.

Of course, Gove being Gove, he didn't do nearly as much wheedling and lickspittling as he should have. His confidential memo to Gallan begins, "May I presume upon the fact that I am older than you by a few months in years and by a few years in service to the Company to make a proposal that I ask you to weigh and consider very carefully."[8] Gallan did not make him a vice president; if Gallan responded at all to Gove, it must have been so galling and humiliating that Gove didn't bother keeping the response in his files.

Gove being Gove, the rejection meant he worked harder to prove to Gallan just how wrong he was to deny him the promotion he felt he was due. Gove being Gove, that hard work redounded to the people beneath him, inexorably pinched under the plate of the press.

Lexicography is a nitpicker's delight: We truck in nested systems, all of which must be consistent or the whole thing falls apart. And Oakes was a good lexicographer, which meant that he was a superb nitpicker. As soon as Godlove's slips began coming in, and though he himself was still recovering from his hospital stay, Oakes began his Gove-mandated review of the definitions. He noted the distinct lack of those supplementary distinctions comparing a color to other colors in the same pocket. Next slip: Do not use the damnable phrase "according to some authorities," because it implies that the dictionary isn't itself an authority—or, indeed, *The Authority*. Next slip: Your use of "to" in the definitions implies a range between two hues, as in "a pale yellow to a moderate orange yellow," so what does "pale to grayish yellow" mean? The meaning of "to" to a lexicographer is very, very specific; Gove wrote a whole damned memo about it.

To Oakes it seemed as though Godlove had put together a defining paradigm after months of consultation with Merriam-Webster and then chucked it out the window in favor of what he thought the ISCC-NBS folks would want him to do. It didn't occur to Oakes that perhaps Godlove's previous stint for the *Second*, more than twenty years ago now, had not been adequate training in the new dark art that was Govian lexicography, or that Godlove was also not blessed with the gift of clairvoyance such that he could read Oakes's mind.

Godlove was, as always, solicitous, but stuck to his guns. He highlights the particular problem of "turquoise green."* This was

* **turquoise green,** *n.* a variable color averaging a light bluish green that is greener and deeper than turquoise (see TURQUOISE 2b) or average aqua green (see AQUA GREEN 1) and bluer and deeper than robin's-egg blue (see ROBIN'S-EGG BLUE 2).

one of those "variable color" definitions that Godlove turned in. He found, as he surveyed all his sources, twenty-five different color chips called "turquoise green"; he had to measure each chip and then give that chip its Munsell notation, which involved juggling about 150 numbers. Then came the tricky stuff: Are all the colors bunched close enough together in the color space that one definition would suffice? If not, how many clusters of turquoise greens are there, and if you made definitions for each one, would those definitions be far enough apart that they would actually read as two different colors to the layperson? Yes, it was somewhat subjective, but scientifically informed nonetheless. So if Godlove chose to use a word like "variable," there was a good (enough) reason, whether it comported with the tight-assedness of Govian style or not. Language was not logic; every system would have its edge cases or exceptions. What was important here was that the science grounding of the ISCC-NBS system be maintained, even if it seemed ridiculous that the color name "turquoise green" was, in actual usage, loosely applied to a range of colors. After all, the consultants of the *Third* lived in the modern world: They knew how imprecise color names could be.

He closed his letter by expressing his "admiration for your excellent grasp of the technical aspects of a highly specialized subject, and for your lucid statements of your understanding of the problems."[9] Nonetheless, he's going to be taking a vacation soon, so don't expect much from him in the meantime.

There was a vacation, yes, but there were plenty of other things keeping Godlove from Oakes's beck and call. Work at the CRL was stressful and isolating, given the huge reduction in staff and the loss of good friends within his department. The ISCC newsletter was increasingly busy: The number of newsletter items that begin with "We have received for review" started reaching double digits. He was diligently working on his book on the history of color, the manuscript of which had already reached about four hundred pages and yet had only covered color from prehistory through the second century BC, but was still tentatively under contract. The previous hospitalizations meant that more breaks were scheduled, doctor's orders.

His work on the *Third* was a source of frustration or confusion for his friends at the NBS, though, who couldn't understand why the lexicographers were second-guessing a system based on scientific principles, and why they were dictating that science move at the pace of publishing, when obviously any scientist knows it's the opposite. When Godlove wrote in 1953 to the NBS colorimetrist Kenneth Kelly to ask if the ISCC-NBS committee had finished all the Munsell notations for the charts—Merriam would need them—Kelly was baffled as to why he couldn't just wait until the NBS had properly reviewed, set, and published the new notations. Maybe Godlove could make an alphabetical list of the measurements he needed and travel from Easton, Pennsylvania, to Washington, D.C., where Kelly could sort through the measurements and read each one to Godlove as they went. "I marvel at the amount of work which you are putting into this revision," Kelly noted.[10] But Godlove did nothing halfway: He told Kelly that all his free time was going into the *Third*.

Six weeks later, right after Thanksgiving, an enormous packet of defining slips arrived from Godlove, along with an explanatory letter. Barring accidents, he says, he'll have no problem finishing the colors in *C* by March 1—the impossible deadline. (This is, frankly, laughable: To achieve this, he'd need to quit his job and do nothing but work full-time on the Merriam defining.) He assured them he had nearly finished the core definitions through the letter *R*, though the supplementary distinctions were still very much in process. That represented a little more than two thousand color names, defined from the ground up, in seven months. If Gove had been paying attention, even he would have been impressed.

Despite pressure from above, within, and without, Oakes settled into his wheelhouse and got to work editing Godlove's definitions. Many of his questions were systemically minded ones. The minutiae of definitions—fine, that is Godlove's domain. But consistency across definitions, across editions, even? That was Oakes's domain. "Crocus" is defined in the *Second* as "saffron yellow" but in the *Third* as "pale to grayish reddish purple," is this correct? Is this slip for "croceus," which *is* defined as "saffron yellow," a typo for "crocus"? You sent a slip for "metallic gray" and one for "metallic grey" that have identical definitions, but we aren't sure if, colorimetrically speaking, "gray" is different from "grey"? As you read the correspondence, you can practically feel, embedded in the paper, Oakes's easy relief in being able to focus on the thing he was best at and ignore the editorial training, the production timeline, the slough of low morale, the disappointed harrumphing of Gove, the feral stink of a cornered animal emanating from the company president's office when it was budget reconciliation time.

Answers were sent promptly on revised slips, and considering the long seesawing relationship that Merriam had with Godlove, they were surprisingly *good*. Revision Godlove was a totally different creature from Specialist Godlove: The definitions of words like "amethyst" had blossomed from the short "dark purple" to the fully fleshed out "a variable color averaging moderate purple (which is) redder and duller than heliotrope (sense 1) or manganese violet, bluer and duller than cobalt violet, darker and slightly stronger than average lilac (sense 1)," and they were now so consistent that all they required was a quick once-over to put them into the house style. The volume of slips was overwhelming: In February alone, Godlove sent in eleven packets of definitions, with about fifty entries per packet. Not that Specialist Godlove didn't make an appearance every now and again. In the midst of defining, he also found the time to write extensive letters on non-definitional issues: the need to begin planning the color plates, the need to revise the article for "color." But tucked into all of these longer letters was the insistence that he was quite overworked and his health was suffering, though he hoped that some rest would stave off a third trip to the hospital. The theme of overwork appeared more and more frequently in these letters—enough that Oakes forwarded one piece of correspondence to Gove immediately after he received it in June and asked if they might be able to get to the color charts before the fall. His reason was succinct: "The man is low."[11]

The next letter Oakes received was from Mrs. Godlove, dated August 14, 1954. It's a stark contrast to the novelettes Oakes was accustomed to receiving from that address:

> Dear Dr. Oakes,
>
> Aug. 8, I.H. had surgery. Appendix were [*sic*] ruptured & gangrenous. He put up a valiant fight until 2:30 A.M. when he passed away.[12]

In such a full life, there was still so much left undone. Godlove had not yet finished his book on the history of color; he had not finished preparing the monumental hundredth issue of the ISCC newsletter under his editorship (which he had dubbed the Jubilee Issue). The ISCC-NBS work was still very much in process, as was all his work at the CRL and several papers he was working on with collaborators in chemistry and optics; he was planning a vacation in the fall. More immediately, the defining for the dictionary of the new scientific age—the defining that was meant to educate generations of Americans on proper color theory and the science of color, while also earning the *Third* accolades and support from the scientific community—was incomplete. All of Oakes's careful cultivation of this consultant was, in the plainest editorial terms, for naught. He had a paradigm for this new scientific way of defining colors, and he had most of the slips in hand, but the scientist upon whom he relied was suddenly gone. No matter how exhausted he might have been, he did right by his consultant. He noted the date on the letter, the very present tense of the tragedy. He sent Godlove's widow, Margaret, a heartfelt if short condolence note. "We who have been favored with a long association with him have suffered a personal loss," Oakes writes. "No one could be more earnest and painstaking in co-operating with our editors."[13]

A few days later, Margaret wrote back to Oakes. She thanked him for his kind note of condolence, noting how thoroughly Godlove enjoyed his association with the company and all the editors he worked with over the years.

She also assured him that she had only a dozen or so defining slips left to complete.

The Golden Girl

Emma Margaret Noss, known later in life as Margaret, was born in 1905 in Pennsylvania, the eldest of three girls born to the Noss family. Her father, Charles, was a pastor in the Reformed Church of America, and her mother, Irene, had been a teacher in Lancaster prior to marrying. (Her parents likely met while they were studying at Franklin & Marshall College, he at the seminary, and she at special Saturday courses for area teachers, where she took a course in algebra.) Education was important to the family: All three girls were sent to Oberlin College, where Margaret graduated in 1927 with a bachelor's degree in chemistry. There are a few pictures of her in the Oberlin yearbook: her senior picture, a dorm photo, a group portrait of the Classical Club. In the club picture, she is smiling broadly, her hair as busy as birds. In her dorm photo, she looks at the camera out of the corner of one eye, arch. In her senior portrait, she looks straight into the camera through the tops of her eyes, one eyebrow barely lifted, a suggestion of a smile settling on her lips, composed and cool. The other graduates look smart, young. She looks like a movie star. She's got your number, mister, and she'll take no guff from the likes of you.

Chemistry was a booming field. According to the 1927 Oberlin yearbook, 17 out of the 262 listed seniors graduated with a degree in chemistry, and 4 of them were women. This seems small, but of all the notable coed schools that turned out chemists during this time, Oberlin had the best gender balance of the era. That's not to say that the four women who graduated with degrees in chemistry had the same career choices that the men did. Typically speaking, and despite the stated need for more chemists in industry, women who were chemists could get jobs as either office assistants or teachers. Margaret opted for the latter, and moved back to her parents' house in Pennsylvania, where she got a job in 1929 teaching general science in the Hollidaysburg High School. Her tenure there gave her a taste of what was in store for her. In her second year of teaching, a Mr. Harry Henshaw was hired to teach alongside her, and she was suddenly made the "health director." This continued until 1932, when Mr. Harry Henshaw was made the physics and chemistry teacher, and she—the chemist—was relegated to teaching health. By 1933, she had evidently soured watching Mr. Harry Henshaw teach her discipline. She applied for and took a job with one of the biggest employers of chemists in the country: E. I. DuPont de Nemours. DuPont was riding the postwar wave that swept wartime tech into the consumer sector. Even though the country was in the midst of a depression, chemistry was still big business, and like many scientists she found a home there. But she wasn't hired as a chemist. She was hired to be a file clerk in the technical laboratory in Deepwater, New Jersey.

Science was the field to be in during the early twentieth century—unless you happened to be a woman. During the late nineteenth century, the sudden incursion of college-educated women into practical science, scientific research, and scientific societies sparked a circle-the-wagons mentality among the men in these groups. Scientific work itself became "sex-typed," as the historian Margaret Rossiter puts it.[1] Women were conscribed to assistant roles, or were made secretaries or clerks, or were relegated to what we'd call the soft sciences, particularly the newly minted field of home economics. If they were lucky enough to have a job, they certainly weren't the head of anything. It didn't matter if they had PhDs, as more and more women of the early twentieth century did; gender stereotypes about women and their motivations ruled the day. The academy was overwhelmingly hostile to women in anything approaching a "hard science"; so few women were tenured professors that it was noteworthy when a woman did anything but lecture in the sciences. Government work was no better. The civil service exam limited women to particular departments, like the USDA or the NBS, because the work at these agencies generally fell in line with common conceptions of what women were supposedly good at: gathering data, measuring things, accurately recording information, certainly not analyzing it. Single women were rarely given jobs with any potential for promotion, since it was assumed that they'd eventually marry, have children, and leave their job to raise the kiddos at home as God intended; married women were no safer, since they could get pregnant and leave at any moment. There was a government hiring freeze of women in 1933, thanks to the Depression, and the Economy Act of 1932 forced women out of government service as soon as they got married.

Women scientists in the 1920s and 1930s were encouraged not to attempt to innovate or become pioneers in fields that were plainly hostile to women, like engineering and chemistry. Instead, they were told to keep an eye open for niche areas where they could tuck themselves away without the fear that men in that field would notice or want those positions. For a woman like Margaret, who had only a bachelor's degree and was unmarried, file clerk at DuPont was the occupational big time.

What we know about her time at DuPont is from snippets she

sent in to her alma mater. By 1936, she was promoted to chief file clerk and had charge of the whole file room. In 1945, she transferred to DuPont's patent department, which, because DuPont was innovating in practically every area of science imaginable, was sizable. (Patent offices were one sure employer of women in the sciences: They needed employees who could read chemical formulas and complex diagrams.) She lists herself alternately as a patent researcher or a patent clerk. Her grandchildren both say that she'd tell them the story of being at DuPont when the company developed nylon, the world's first fully synthetic fiber, and how they had to change the formula for the commercial market because the original nylon didn't snag and run, and how could you sustain an entire industry on an indestructible consumer good you'd need to buy only once?* It was here that she met I.H. They became friends, and their friendship later grew into a courtship that must have been long-distance for a few years, since I.H. moved to Easton, Pennsylvania, to take his job at General Aniline in 1943. On August 6, 1949, they were married in Steelton, Pennsylvania. Her cousin Christopher Noss was the officiant, and the witnesses were her sister Mary and I.H.'s son, Terry. Her next alumni survey to Oberlin is posted from Easton, and she lists her career right after her marriage as "homemaker."[2]

It's true she was no longer working at DuPont, and it's true that she was now busy with the duties of a housewife. But think back to her graduation picture. That eyebrow, that little smile, that leveled and leveling look. Give up a career? *We'll see about that.*

It was not an unusual thing through the nineteenth and twentieth centuries for the wives of scientists to be active in their husbands'

* I couldn't confirm this in the DuPont archives, but I can't imagine they'd happily announce this news to the pantyhose wearers of the world.

scientific pursuits. With the right partner, a scientifically minded woman could find something intellectually stimulating in the marriage, though she'd have to be content with being a lab assistant, the second (unnamed, uncredited) fiddle. Fortunately for Margaret, she married a pretty decent scientist who was known for being a collaborator—and also struggled with some workaholism. By the time they married, I.H. was the head of one division of physical chemistry at the CRL and in leadership in four different professional organizations, including the ISCC, and he was trying to persuade Merriam-Webster to let him take on even more work.

The ISCC newsletter was Margaret's entry point into the world of color. The newsletters had grown expansive under I.H.'s editorship. Initially just reports on member societies and the minutiae of the ISCC Executive Committee, I.H. sought to make them the *Harper's Bazaar* of the color societies. In addition to the usual reports on member societies, conferences, and publications that touched on color, I.H.'s newsletters included reports and reviews on news stories that mentioned color, riddles about color perception, not terribly funny jokes about color that looked as if they might have been nicked from *Reader's Digest* or *The Saturday Evening Post,* and extensive disquisitions on the place of color in history. Under I.H.'s editorship, the newsletters went from generic news items on color in art, science, and industry to avuncular and professorial chitchat that just so happened to center on color. Gone were the dry reports that subgroups of color specialists were organizing geographically. In their place, an over-the-fence neighborliness without stiffness or formality. "As for Hollywood," I.H. writes in 1945, "we hear that a [color-science] group there is a-borning. Let us know when the christening is, so that it can be properly announced in the pages of the News Letter."[3] As the ISCC's ranks grew, and particularly as those ranks included more laypeople and fewer scientists, more articles on the general problems of color—

matching, standards, definitions, naming, observing—appeared in the newsletter. Opinions of "The Editor" were sprinkled liberally throughout all news items. It's no surprise, then, that when I.H. died, many of the ISCC members felt as though they knew him intimately: He was all over the newsletter.

The work involved in creating the newsletters was eye goggling. The newsletters averaged fourteen single-spaced typewritten pages, but some of them were upwards of thirty-five pages long. Each item that came in had to be logged and evaluated for inclusion, then edited down into the third-person voice used in the newsletter, retyped into the newsletter format, and filed away for safekeeping. A cordial and gracious editor also keeps up with the inevitable correspondence that results from running a newsletter: correction reports, response letters, advertisements, cranks who just want your time and attention. Mr. Harry Henshaw would have dusted his hands and proclaimed it was perfect women's work.

There was also likely another inducement for Margaret to get involved in as much of I.H.'s work as possible. In April 1949, just a few months before she and I.H. were married, I.H. was hospitalized. He wrote candidly about it in the May 1949 newsletter:

> Dr. Godlove, editor of the News Letter and present chairman of the Council, has been ill since early April. For a long time he has been pushing himself on the many projects that interest him, and Nature has finally caught up with him. Fortunately, although exhaustion made it necessary to hospitalize him and provide three shifts of nurses, all tests so far have proved normal. We hope for him a period of rest and convalescence that will soon bring him back to his usual vigor.

Nonspecialists probably imagine producing dyes as something akin to high school chemistry: Put on your safety glasses, mix some things together, maybe the worst that happens is you stain the workbench and your chem teacher yells at you. But dye manufacture was dangerous, even in the lab-controlled conditions of the twentieth century—and not just because of the chemical weapons derived from its by-products. One 1920 report published in the *American Journal of Public Health* lists, as possible injuries at a dye works, chemical burns from exposure to strong acids, pulmonary edema from inhalation of toxic fumes, cyanosis, and poisoning.*[4] More modern studies have noted that chemists who synthesize dyestuffs can be routinely exposed to benzidine, trichloroethylene, acetic acid, and biphenyl, all of which have been linked to various cancers in humans. It was so dangerous, in fact, that DuPont's bellwether Haskell Laboratory of Industrial Toxicology, which later focused on consumer safety testing, was formed in 1935 because a huge number of the DuPont employees at the Deepwater dye works suddenly came down with fatal bladder cancer. And just to top it all off, there were also occasional explosions:

> In 1945 an explosion blew out windows and wrecked equipment in the research building. Three employees were injured: Fred Albinson, a chemist; George Keck, a company fireman[;] and Kenneth Town, a lab assistant. My wife, the former Doris Dean, who worked in the laboratory, clearly remembers the incident as well as myself. The blast buckled the ceiling of the basement and the floor of the first floor, knocking over and destroying analytical balances where Drs. Louis Waldbauer and Larry Hallet worked and near the area where Dr. Isaac H. Godlove worked as head of the physical chemistry department.[5]

* "The earliest symptoms we get in the workmen are headaches, giddiness and a 'down and out feeling,' as the men describe it, and then upon careful examination a certain amount of cyanosis will be discovered. . . . Rest and fresh air are very essential in the treatment."

If I.H. suffered, then, from "poor nerves," at least he had a good reason.*

Regardless, he was back at it for the July newsletter, one month before his wedding, though still a bit subdued. He signed his name to only one one-page addendum to a book review, and included only two pages of his "History of Color" manuscript. But marriage must have suited him: The September newsletter included a *Reader's Digest* extract on bright colors, a seven-page bibliography of articles and books on color, and two pages of color reports from the TCCA, condensed from sixteen separate bulletins of "single-spaced typing, packed full of palatable meat" (the best kind).[6]

Though Margaret never got a shout-out from I.H. while he was the editor of the newsletter, her pencil marks were all over it—literally. Copies of I.H.'s newsletter entries on color are annotated in the margins with her neat handwriting; his manuscript on the history of color is copyedited in two hands, his and Margaret's; the running bibliography that appeared in the ISCC newsletter was managed by Margaret, even if I.H. and others gave her the citations. Margaret's work in compiling the newsletters familiarized her with the major players in the color world of the 1950s. She became friends with his CRL co-workers Hugh Davidson and Henry Hemmendinger (to whom I.H. had dished about the pretty girl at DuPont before said pretty girl became Mrs. I. H. Godlove); she was in constant communication with Dorothy Nickerson, the USDA colorimetrist who gave the ISCC-NBS standard its name, and Judd, the NBS colorimetrist, both of whom worked closely with I.H. on the ISCC newsletter committee and a number of I.H.'s side hustles; she typed up ISCC-related and personal correspondence for I.H. and, in doing so, had to make sure she compre-

* Maybe he wrote about the explosion for the CRL monthly employee newsletter, *The Anilog:* He's listed as a features reporter in the September 1946 edition, because of course he is.

hended what, exactly, I.H. was supposed to be saying. She kept his files; she recorded research. She was a smart and curious woman, and one whose history uniquely equipped her to carry on I.H.'s work.

As with many women who are consigned to being "the wife of," Margaret's contributions to I.H.'s work are primarily read in between the lines of I.H.'s own papers kept after his death. Some of them were deposited with the ISCC archives, but Margaret kept an assortment of his working papers and passed them along to I.H.'s son, Terry. Certainly his manuscript and the raw materials for the newsletters are evidence of some work on her part, but we get a fuller picture from the sympathy letters people sent her.

Condolence notes from everyone in the color field expressed shock and concern for I.H.'s sudden death, and shared their remembrances of him as generous and charming, someone who made both the ISCC newsletter and the field of color a more welcoming place to all: "a fine person and splendid scientist," as one of his former co-workers wrote.[7] Those sorts of sentiments of high esteem are to be expected in condolence notes from co-workers and friends; what's unexpected are all the commendations of Margaret. "We all saw what you did to and for I.H.," wrote Miriam Hemmendinger, Henry's wife.[8] Several people commented on I.H.'s precarious health ("I know I.H. was never strong," writes a friend from the DuPont days[9]) but noted that his death had come as a shock because he seemed so much better than he had been—because of Margaret. Many thanked her for all that she did in supporting him and his work. Kenneth Kelly's condolence letter finishes with, "I look forward to our visits at future meetings and join your many friends in urging that you do not deny us this pleasure."[10] Indeed, she was so much a fixture in the color world that when the *Journal of the Optical Society of America* wrote its memorial of I.H., the first paragraph mentioned how much the society had come to enjoy *Margaret* and I.H.'s presence at meetings, and the article was accompanied by a pen sketch of Margaret and I.H. at the last OSA meeting, Margaret in the foreground.[11] The scientists who knew I.H.'s work most intimately knew what they owed her. "That you shared too in this same esteem is evident in a comment—'what a fine couple!'" wrote one of the officers of the American Association of Textile Chemists and Colorists (whose letterhead sensibly shortened the name of the association to AATCC).[12]

The most explicit letter of condolence is from one of her newsletter cohorts, Dorothy Nickerson. "You have been wonderful to—and for—I.H. these last few years when it has been so hard. He was a remarkable person in a very great many ways, yet none-the-less (and perhaps more, because of it) you have gone through much for him," she writes a few days after I.H.'s death. Then down to business:

> Whatever do you think we should do [about the newsletter]? Should we continue with the Jubilee Issue and make it a memorial issue? It was so much a part of I.H. that I want it to be completed as he would have wished, and to do that I need your advice.
>
> Are you completely worn out—or would, or could you—carry out and complete the next few issues as you think he would have wanted them? I'd like to know what you think should be done.[13]

The Jubilee Issue went on as planned, in no small part due to Margaret and Dorothy's work. It was a stunner. If readers had no idea how far-reaching color measurement, study, and standardization reached, they got a taste of it looking at the table of contents. Progress reports came in on color measurement, color instruments, color ordering systems, dyes, synthetic dyes, color and ceramics, color and glass, color and plastics, color and oils, color and leather, color in the military, color in one particular branch of the military ("Naval Affairs"), the psychology of color, the biology of color, color in photography, color in color television, color in art, which was distinguished from color in the graphic arts, which also was distinguished from color print, the latest fad for using colors, colors in home furnishings, which was different from colors used in the home, which was different from colors used in architecture, which was different from the promotional use of colors. The issue ended up being forty-four single-spaced typewritten pages, bound in a color-printed cover designed by I.H. himself.

Dorothy Nickerson was the stand-in editor for the Jubilee Issue, and it is in her foreword that Margaret is finally given some props. Her foreword begins with an explanation of what the Jubilee Issue was intended to be and notes that I.H. had planned it out so thoroughly that all that needed to be done was to follow through. "This has been done in large part by his wife, Margaret Noss Godlove, to whom we express our sincere thanks."[14] Margaret also warrants another mention in the foreword—as the person who, since marrying I.H., prepared most of the ISCC newsletter material, gathered and acted on editorial comments from the other editors, and then sent the whole shebang off to Mimeoform Service in Washington, D.C., for printing and distribution.

That was apparently the last newsletter issue that Margaret had anything to do with. But unlike the widows of other ISCC members, who blurred and drifted away, Margaret stuck around. In May 1955, she was received as a member of the ISCC on her own, despite her lack of academic credentials, or her work in industry.

In many of the condolence notes sent to her after I.H.'s death, some of her family encouraged her to move closer, to close up her life in Easton—clean break, find folks to take care of her, that sort of thing. Another friend was of another opinion. "How wise you are to stay right there and continue your life in much the same pattern," she writes. "Then when you are ready, take a job. Keeping busy is to be your greatest blessing."[15]

Margaret, ever resourceful, was way ahead of her.

Pink Collar

The Merriam-Webster building in Springfield, Massachusetts, is part working office, part museum. The entryway is chockablock with historical ephemera—a giant wooden case for the 1912 World's Fair that houses old dictionaries, the old G. & C. Merriam Co. wooden sign that used to hang outside George and Charles's printing office, even an end table with silk flowers and a rotary phone (only for effect: The phone's not tied into a line). But venture past the entryway and head upstairs and you'll find yourself in a tidy warren of cubicles covered with plain-weave fabric in Ambiance, Mute, or Hush. Turn about-face and you find the historical heart of Merriam-Webster: the citation files.

The citation files are where all the historical evidence for entries is stored. They are long, ruby card catalogs, about four feet high and twenty feet long; open any drawer, select any word, and you get a narrative wiring diagram for most entries in the major Merriam-Webster dictionaries printed during the twentieth century. Inside you'll find the citations, or cits, those bits of language in the wild scooped up in context and ferreted away for some future defining event. But filed in and among the cits are other notes that are the baling twine and glue of every entry. There are cards called pinks (so named for the color) that are in-house notes about the entry, and then there are "buffs" (ditto), which contain the definitions, both finished and heavily edited, for every entry. When lexicographers talk about the process of writing a dictionary entry, they tend to give most of the airtime to the collected citations. They're the raw material used in orienting and crafting a definition, after all: the whole reason a dictionary exists.

Those pinks and annotated buffs tell a fascinating story of how these definitions came to be. There are, as mentioned earlier, the marked-up definitions from the *Second*. There are new definitions consultants have sent in for the *Third,* or which were written by in-house editors for the Big Book. There are notes back and forth between editors on how to handle the smallest jot and tittle of each definition. All of these are stamped with the editor's name and date so any curious person can construct a historical timeline of how an entry progressed.

In the end, the tell was the math. The color definitions for the *Third* were sent into the office in an overwhelming flood. First, there are the responses to the pasteups from the *Second:* 2,170 of them in all, many of them either being junked or needing a full revision. Just sorting through them and determining which ones are in the authorities cited, or which ones have significant historical or current use (and so will be entered in the *Third*), is a time-consuming and utterly thankless task requiring extensive research and cross-referencing. Then you have to move on to the actual defining. There are more than three thousand colors in the *Third,* all of them requiring the thoughtful ministrations of the color specialist, the same one who repeatedly stated that he had a demanding day job, a raftful of societies he's leading, a book in the works, articles to publish, and had been hospitalized several times for overwork. The odds are not in Webster's favor, and yet all the defining slips come in between August 1953 and March 1954.

I want to belabor this point because it's an impossible feat. I.H. completed three thousand or so new entries in seven months, with a good fraction of those entries having multiple color definitions. That works out to about a hundred definitions a week, or twenty definitions a day. For the blessedly uninitiated, that is a blistering pace even for the best professional lexicographer working full time. Each of those definitions couldn't just be dashed off; it needed to be fully researched, which involved taking measurements of multiple color chips, mapping them onto the Munsell system, translating their position in the Munsell color solid into words, and then typing that definition up on a buff and stamping it with the date for posterity's sake. Even with a formula for defining and the prework of measuring and ordering done, there will still be entries that you have to poke at and knead and slap onto the counter a few times before they begin to take shape. It's one thing to dash off a definition for "Yale blue"; it's another thing to dash off a definition for "blue."

Oakes knew this: He was a veteran definer, and he was herding other consultants, many of whom took half a dozen years to produce a few thousand entries. He also had monthly reports from I.H. that were less than promising—under the weather last month, a break for vacation two months ago, very busy with the society and newsletters. Yet here were the buffs—hundreds and hundreds of them. How did Godlove do it?

"Chrome green" is the first alphabetical entry that lays bare the mechanism behind the miracle. The file contains the multipage entry submitted by Godlove in February 1954:

chrome green

1 any of several colors, all dark yellowish greens, of mixtures of Prussian blue and chrome yellow with varying amounts of white pigment. Deep chrome green is yellower and duller than holly

green (sense 1) or golf green; yellower and less strong than average hunter green — called also chrome green, cinnabar green, Milori green, silk green, zinc green, zinnober green. Medium chrome green is yellower and duller than holly green (sense 1) or golf green; yellow and less strong than average hunter green; stronger and slightly lighter than deep chrome green. Light chrome green is yellower and less strong than holly green (sense 1); yellower and darker than golf green; yellower, lighter, and stronger than average hunter green; lighter and stronger than deep or medium chrome green — called also navy green, Windsor green

2 the color viridian; the color of chromium hydroxide or hydrated chromium oxide (the pigment commonly called chrome green in the British Isles and transparent chromium oxide in America)

3 the color of chromium oxide, commonly called opaque chromium oxide[1]

To the trained and discerning eye, this entry is a hot mess. Oakes and Godlove had already come up with a defining style for these colors, and yet here was a "new" definition of "chrome green" that was just a warmed-over slapped-up version of the entry as it stood in the *Second*—that's the whole "mixtures of Prussian blue" bit—only this time it's lavishly *and* inconsistently punctuated. Had Godlove been an in-house editor and subject to Gove's exacting standards, that rampant use of periods and semicolons would have gotten him ejected from the office immediately (but very quietly). The only part of this entry that almost adhered to the agreed-upon defining standard for colors was the long part of the first definition that covered deep, medium, and light chrome green.

The entry was revised by Oakes, then Gove, then Oakes again in 1955. The revisions leave almost nothing of the original: Gove's heavy red pencil crosses out almost all five pages of Godlove's professorial definition. But the problem was that Godlove included some nervous-making cross-references to pigments, which were not, dictionary-wise, Godlove's purview. It was up to Oakes to combine all the separate definitions from all fields—chemistry, pigments, dyes, color—into one cohesive entry.

It's an adventure fraught with peril.* Anytime a general editor has to revise or copyedit an entry that involves highly technical (usually too technical) language from a consultant, there's the risk of simplifying the entry into error. Fortunately, Oakes was a good compiler. He put all his collation-related thoughts down on one pink, and this is where the man behind the chrome green curtain is revealed. He has three questions about "chrome green," and all begin with "Query to Mrs. Godlove June 15, '55."[2]

His letter to Margaret asks more complex questions than you'd dare ask a mere stenographer. That second definition: Is the color of chromium hydroxide the same as viridian, or are these two separate colors that should have been made into different definitions? Is that "British chrome green = US transparent chromium oxide" note in the second definition implying that the color of those things—pigments, are they?—is the same? There's another slip for "chromium green" (the color) in the batch that implies it's the same as "chrome green" (the color)—is it the same pigment? He's not asking her to send duplicate slips for entries, or to see if I.H. left any clarifying notes in his files. In fact, *Dr.* Godlove isn't mentioned once in Oakes's letter.

It wasn't as if Oakes had no other recourse. By 1955, Margaret had referred some of the more complex questions regarding color charts and the lengthy article at the entry for "color" to Dorothy Nickerson and Deane Judd. If he felt that Margaret wouldn't be able to give him the clear answers he needed, he could have cer-

* For very small potatoes yet still stressful values of both "adventure" and "peril."

tainly asked them. But Oakes didn't ask the other colorimetrists. He asked Margaret.

Her answer is thorough. "The color viridian" and the color that is characteristic of chromium hydroxide are the same; the British chrome green refers only to the pigment and not to the color, because the chrome greens are in the dark yellowish-green pocket, and transparent chromium oxide is in the moderate or strong green pocket. And that slip for "chromium green" sense 1 refers only to the color, not the pigment. She also apologizes for the delay in responding so she could research the questions further. The delay was only six days.

It was I.H. himself who got the ball rolling in Margaret's direction. In a letter to Oakes dated May 29, 1953, I.H. lays out his difficulty: He's been ill and his recovery has been somewhat touch and go; he's needed at General Aniline, as well as at the ISCC and as the color chairman of the AATCC. He notes that General Aniline will only let him work on the color definitions provided that it doesn't interfere with his current work or result in a downturn of his health, and, he says candidly, he doesn't frankly know whether he'll be able to take up the color work on those terms. But he has an idea that should safeguard a quick completion of the work, such that even if he has to hand it off to someone else, it will be nearly completed. On the second page, buried in and among complaints of Granville's matte and glossy chips being miscalibrated and the woes of glossy materials for the colorimetrist, he slips in the suggestion that Margaret be paid $1 an hour for "stenographic work," which will help him at least get a start on the color defining. This suggestion is sandwiched between a casual note explaining that "Mrs. Godlove has recently helped me enough on this [the defining for the *Third*] and other color work" and a somewhat embarrassed admission that he's still not feeling well enough to work at full speed.

It's a deft bit of word craft. I.H. knows that Merriam-Webster wants the top color expert in the field writing their definitions. Being chosen by the top dictionary company in America to write definitions for what will be the crème de la crème of modern dictionaries is an acknowledgment of your unerring mastery of the subject. It earns you a little capital. The partnership would be good for Merriam-Webster as well: To be on the bleeding edge of new color terminology would be not just forward-thinking but economical; Merriam wouldn't need to update their color definitions for a while, so no need to hire another color consultant in ten years. And I.H. knows from his association with Merriam-Webster in the 1930s that they didn't object to his having an assistant help type up his slips, so they certainly wouldn't deny his request to have Margaret compensated for stenographic work. But it's also clear that I.H. knows how much of the defining work Margaret will be doing, which perhaps is why he feels compelled to reassure Oakes that "they" would be working on it together as time permits, and that Margaret has helped him already with other color work. He won't come right out and say, "Margaret will be doing the defining," lest Merriam dump them so far along in the process. But I.H.'s regard for her and her know-how was profound. He could have referred the work to Deane Judd or Dorothy Nickerson. He could have thrown his back into it and done as much of the work himself as possible; after all, *he* didn't know he was going to die in 1954. He could have used Margaret and her work and just not acknowledged it to anyone at Merriam-Webster—which was, frankly, the established mode for most of the special editors and consultants, with their platoons of uncredited grad students and secretaries. But he refused. Not only did he love Margaret, but he viewed her absolutely as equal—equally as smart, equally as important.

From this point on, I.H. drops all pretense, and Margaret's name begins appearing regularly in I.H.'s letters: "Mrs. Godlove has alphabetized about 100 of the [NBS] report's 267 classes"; "Mrs.

Godlove has completed the indexing"; "Mrs. Godlove have [*sic*] put a great deal of time on the color definitions and have [*sic*] virtually completed 77 of the 167 classes of the ISCC-NBS report."[3] In June 1953, I.H. made it clear that he had come up with the basic defining structure for the definitions, and that it was up to Margaret to use this structure to go through the colors in the ISCC-NBS report and define them. The "haves" in this last letter are interesting grammatical slips, plural where it should be singular: It's clear that I.H. thinks of the work as being completed by both him and Margaret, working in tandem. But the sentence begins only with "Mrs. Godlove."

The signs, however, were there for those with eyes to see. The number of times that I.H. had used "we" when talking of taking measurements of chips, or organizing the defining principles, is eye-opening; the number of times that he had mentioned Margaret in proximity to the Merriam work shouldn't have made his May 1953 letter much of a surprise. "Mrs. Godlove" was a regular character in his letters to Oakes, taking measurements, indexing, organizing. It was clear that I.H. trusted her; who was Oakes to gainsay that? There was no pushback from Merriam-Webster on I.H.'s decision.*

The pinks and buffs in the file aren't enough to get the full story; when they are lined up with the correspondence, it's clear just how much Margaret had to have been doing. While the actual writing of definitions progressed in a straight line from May 1953, when Margaret began working on the definitions, to March 1954, when the definitions were supposedly all turned in, I.H.'s letters are still a slalom course full of swerves from one new idea to the next. It

* It has been suggested to me that maybe Oakes and Gove had a face-to-face conversation about this unusual arrangement, rather than setting down their thoughts in a memo. Silly person: Lexicographers always set their thoughts about dictionaries down in a memo, and they do everything they can to avoid any face-to-face conversations in the office, period.

might be a good idea, I.H. writes, to change the defining formula to begin with "a red/blue/yellow/green color," even though this very suggestion was one of the first things Oakes jettisoned for the *Third* (June 1953). How about starting the definitions with "according to some authorities" when the color name was used of a broad range of colors (July 1953, *no*)? Heads up, I'm changing how I'm defining many terms so that "light [color name]" and "dark [color name]" will be completely self-explanatory (August 1953, *please stop*). What about a diagram of color terms arranged into the categories of "physical," "psychophysical," and "psychological," just an extra page in the dictionary is all that should be needed (November 1953, *oh my sweet Lord*)? Oakes summed up his thoughts on the matter at the top of a pink to Gove: "Endless fussing with color defs."[4] Yet despite the extensive single-spaced meanderings, buffs poured into the office.

I.H. didn't come up with the agreed-upon formula for the color definitions until June 1953. Margaret had, by then, indexed all the colors to be entered according to the 267 pockets they belonged to. She had accurately measured all the colors to be used, converted them into the Munsell notation, and organized them by hue, value, and chroma. The core definitions—"a vivid red"—were based on Margaret's work organizing the color names into each of the classes. But it's not always clear who wrote the secondary distinctions. In one update to the Merriam offices, I.H. says he's been dictating the distinctions to Margaret; in other updates, he says that Margaret has completed *X* number of definitions. I.H. has created the template for these definitions, but it looks as if Margaret moves forward and does the drudge-level work of defining when I.H. is too busy—which is often. It's also clear that many of the definitions must have been written well before they were date-stamped; the date stamps on the buffs for the color definitions coincide better with the date on which they were sent to the office, not when they were written. This would account for some of the earlier definitions, like "chrome green," being in several defining styles at once: Margaret had to fly blind.

Though I.H. had a plan for using the most commonly known colors as comparison colors, he doesn't explain how he's determining which colors are most "commonly known" to the layperson, which means there's a little latitude in how these definitions are actually oriented. And latitude is where a lexicographer can be creative:

> **teal duck** *noun* . . . **2 :** a dark greenish blue that is bluer and duller than average teal, averaging teal blue, drake, or duckling

There is something delightful and just *right* in the color definition for "teal duck" being oriented to three other colors named for waterfowl (teal, drake, and duckling*). The same thing happens at "buttercup," defined in reference to other yellow flowers:

> a variable color averaging a vivid yellow that is redder and deeper than dandelion (see DANDELION 3b) or goldenrod (see GOLDENROD 2a)

This seems like a sly joke, but it's actually a decent decision on the part of an amateur lexicographer and experienced color scientist. Margaret, more than I.H. and more than Merriam, straddled the world of the specialist and the generalist. She respected that they had to somehow communicate small color differences to people whose only source of color information was what they read in the Sunday paper and what they saw in the Sears catalog—and,

* Which, I was surprised to discover, is *not* the soft lemon yellow of newly hatched ducklings, but a color that is, in nearly every possible way, exactly its opposite: a dark, saturated blue. I don't know why I'm still surprised by these sign-signifier incongruities given the whole "sea pink" hoopla, but here we are.

as a woman and thus the target demographic of most marketing, she would have been much more familiar with the way that magazines, catalogs, and advertisements grouped and described colors. She understood that those intrinsic colors, though perhaps not the most common colors by I.H.'s estimation, were the ones we had the strongest associations with and, according to the research she proofread for the newsletters, were surprisingly stable in terms of the color we associated with the name.

Margaret had good lexicographical instincts, and this was a welcome change to I.H.'s constant stratospheric waffling. Anytime Oakes had asked I.H. a question about a definition, he had gotten a lot of academic convolutions in return; with Margaret, it was simple and straightforward answers. Which sense of "lemon yellow" should we put the "called also" at? I.H. would have sent in a scholarly article's worth of discussion in response. Margaret's response: definition 2. She refuses to take on the definitions for the basic color names, like "red" and "blue," saying it would be "quite unfair"—they're beyond her own abilities—but Oakes is no dummy. "You are to be commended for undertaking to carry on and Dr. Godlove's training is to be credited with your being able to," Oakes writes her.[5]

She was still being paid the stenographic wage of $1 per hour.

The buffing out of women and their discoveries from the general narrative is indisputable. Differentiated sex chromosomes, nuclear fission, wireless technology, the modern refrigerator, windshield wipers, Kevlar—all these were discovered or invented by women, yet how many people can name the discoverer or inventor? (You might smugly proclaim that Hedy Lamarr invented wireless tech, though chances are good you remember that because Hedy Lamarr was a bombshell and an actress, and only upon reading more about her did you discover she was the mother of Wi-Fi.) Women's work went uncredited or was appropriated by male colleagues, and it was rarely corrected in the moment. The same year that Margaret Godlove was given full rein over the color defining for the *Third,* Rosalind Franklin, running a parallel track with three other scientists in London, determined that the structure of DNA had to be helical based on X-ray diffraction photos she had taken of the molecule. Franklin's research was credited to James Watson, Francis Crick, and Maurice Wilkins, all of whom were awarded the Nobel Prize for "their" discovery.

But color was one of those aforementioned soft sciences where women could make a place for themselves. Studies of gender differences from the late nineteenth and early twentieth centuries often claimed that women were more sensitive to color and more detail oriented in general; other studies were starting to link hereditary color blindness to men. Women were therefore thought to have better—or more secure—color perception than men. And because color touched nearly every science in some way, a surprising number of women were involved in some of the foundational work of color standardization. The DuPont and General Aniline labs both employed women as technical and lab assistants, making women an integral part of industrial color manufacture. Women at the NBS and other agencies learned how to operate the scientific instruments like spectrophotometers and colorimeters that measured color, putting them right in the midst of color standardization. The Munsell Color Company, small as it was, was owned by a man but managed by a woman: All of the details regarding the production of the various color standards, books, and educational materials that the Munsell Color Company produced were under her purview. The heyday of color design also meant women were called upon for their (supposed) intuition as interior decorators and color consultants in architecture and urban planning. And, of course, women had always been involved on the color side of fashion: Margaret Hayden Rorke was just one of many women who were involved in color forecasting for various companies.

It's not entirely surprising, then, that women also co-authored a number of important color systems, standards, or dictionaries in the twentieth century. The ISCC-NBS standards are themselves based on thirteen different sources, from postage-stamp color-name standards to previous color-name dictionaries and atlases. These thirteen sources were chosen by the ISCC and the NBS as the best-known and most scientifically respectable color standards of the day. Six of these sources were either written by women or adapted by a woman for use by the ISCC. That includes the Plochere Color System, created by Gladys Plochere and her husband, Gustave, for use in their design business;* the TCCA color standards, pulled together and overseen by Margaret Hayden Rorke; and the *Descriptive Color Names Dictionary,* written by the colorist Walter Granville, the commercial color consultant Helen D. Taylor, and the commercial color consultant and designer Lucille Knoche. But even in a field as open as color, there are still erasures: This last color-names dictionary is commonly known as Granville in the literature.

It was much the same in lexicography, truth be told: Women were essential to the creation of dictionaries, but so few of them are remembered for their work, nor were they acknowledged or rewarded for it in their own time. Take, for instance, the beleaguered Anne M. Driscoll. Miss Driscoll, as she was called in the office, was hired in 1937 as an editorial assistant, one of the many Smith College students recruited to work for Merriam by Neilson. She began, as did all editorial assistants, with lower-level tasks: alphabetizing, proofreading. She proved herself a capable

* The Plochere Color System is still a family business: Gladys and Gustave's granddaughter, Michelle, runs it now and still offers color consulting services.

editor and was promoted up the editorial ranks, becoming what was essentially a production editor. By the time Gove became the managing editor of the *Third* in 1951, she was an assistant editor who had helped train and supervise the readers for the citation collection program alongside Oakes, defined and provided all the abbreviations for *Webster's New Collegiate Dictionary,* and was, according to several editors who were hired during the *Third,* the acting managing editor. She was invited to be on the "small council" that helped set editorial policies, and her initials appear on memos about defining technique, how to style numbers in definitions, how to handle compounds, how to handle variant spellings.

Yet despite the actual editorial work she did, she was slighted by Gove. She was invited to take part in the editorial board meetings, but only as a secretary; she was, I will say, a fantastic stenographer, but it's not as though there was a dearth of clerical workers at Merriam at the time and she was the only option. Her personalized date stamp is all over pinks and notes from the *Third.* She stuck it out through staff turnover and fatigue, low pay and high stress, diligently checking cross-references, querying, editing, arranging for proofs, revising proofs, staying late. When the powers that be decided to officially give Gove the title of editor in chief rather than calling him the managing editor in 1961, the assumption in the office, then, was that Gove would go ahead and select the two associate editors who had gone above and beyond in the production of the *Third* and give them the titles of general editor and managing editor on the masthead of the *Third.* He decided to do no such thing; Driscoll, who had been doing all the work that a managing editor would have, quit in disgust upon hearing the news in 1961. She was immortalized in a parody puppet show that Gove's wife, Grace, wrote about the creation of the *Third* and that was produced at a party for the editors after the *Third* was released. Driscoll makes an appearance in the show alongside Gove, Gallan, and H. Bosley Woolf (the man who would take over for Gove upon his retirement in 1972). Knowing only what her husband told her about how the office operated, Grace wrote Driscoll to be an irritating, near-hysterical meddler who only enters the scene to deliver sheaves of words from the consultants. When Driscoll's character exits, the three men joke that she needs a man, or a lover, or to take up golf. As she flits in and out, the three men are crouched over a cauldron where the *Third* is bubbling away, each stirring and murmuring incantations, the only ones hard at work producing the dictionary. When the Big Book finally appears in a puff of smoke, it's Gove, Gallan, and Woolf who get to sing, "We are lexicographers / who made a big book / by damn."[6] Driscoll is offstage. Her sudden departure in Grace's play, never to be heard from again, mirrors what happened in real life. She's listed in the *Third* as an associate editor and nothing more.

The backbone work that goes into creating a dictionary is often intentionally swept under the rug in favor of a very nerdy *The Old Man and the Sea* narrative. One solitary lexicographer adrift on an ocean of language, pulled farther and farther from the shore as their catch (say, the verb "do") tries to escape; attacked by the distractions of office coffee and customer email; dragging their skeletal definition back to shore, lashed to their little boat. Lexicographers get positively swoony about the mythic struggle that defining can be, and we all like our epic tales to be pared down to one hero and one villain, one epic contest. And so we talk about Webster's 1828, or Johnson's *Dictionary;* we talk about Gove's innovations in the *Third,* or Neilson's expansive scholarship on the *Second*. But that effectively erases that every part of lexicography has to be a group effort, from the ground floor up: Using the language, collecting it, organizing it, analyzing it, all require more than one person. There's a reason that modern dictionaries don't let the lexicographer who drafted the entry sign it: Even the process of whipping a drafted definition into shape is staff-driven, not one that relies solely on that nerdy fisher of words. More rests on the editor, typist, or proofreader than rests on the initial lexicographer, since a dictionary is judged by what makes it into print rather than what doesn't. The importance of this work and who does it is borne out in sheer numbers. A photo of the Merriam-Webster staff taken in 1948 shows about seventy-five people jammed together on the front steps of the building on Federal Street, nine rows deep. About two-thirds of the staff are women.

Lindsay Rose Russell, a professor of English at the University of Illinois Urbana-Champaign who has researched the role of women in writing dictionaries, sums it up: "Dictionary making is not a romance, even between a man and a language. It is a rhetorical and material production that . . . was underwritten by women's labors even as it foreclosed on their participations in dictionary production and content."[7]

Over a lunch of *ropa vieja* and plantains in a Maryland suburb, I met Terry Godlove Jr. and Karen Malley, I.H.'s grandchildren and Margaret's step-grandchildren, to ask them what they remembered of I.H. and Margaret. It's an awkward thing to have a complete stranger contact you out of the blue and tell you that they're writing a book about your grandparents: a bit like having a photographer knock at your front door and ask to come in and document your underwear drawer. If they were wary, they didn't show it. They were as generous as their grandparents were reputed to be.

Neither Terry Jr. nor Karen met I.H.; he had died before they were born. What they knew of him, they heard secondhand. He was scientific and precise, but generous, a man with no ego. But they both knew Margaret well. Margaret had adopted Terry when she and I.H. got married, though Terry was an adult, and she was very involved in Terry Jr.'s and Karen's lives as they grew up.

Terry Jr. remembers Margaret best not as a color scientist but as a veterinary assistant. After her color work was done, she worked for a large-animal vet in Easton whose office was in the same building that she lived in on Spring Garden Street. She had grown up on a farm and so it was a return to her roots; she assisted the vet occasionally on farm calls and brought her grandson along. He doesn't recall her doing color work much past 1963 or 1964.

What they both remembered was visiting her and playing with her copy of the *Munsell Book of Color,* or flipping through I.H.'s collection of art prints and books collected from libraries and museums—ancient art, medieval art, modern art. Margaret didn't talk about the color world with her grandkids, though, and neither of them had any idea that she might have been involved in I.H.'s defining work at Merriam. Terry says, "She did grandmotherly things. She made a tremendous angel food cake."

Margaret's angel food cake was, according to Karen, "to die for." Crusty on top, moist and luscious on the inside. Karen said she's determined to replicate it. "It's a worthy quest," Terry responded. The angel food cake, like the color defining, was just one part of a chemist's life. "Part of her training was chemistry as applied to food science," Karen explained. "She really understood about egg whites, and the temperature of things, and the kind of pan, and the quality of the ingredients." Margaret left extensive notes in her angel food cake recipe, notes that mark her as a methodical and scientifically minded woman. It runs in the family: Terry Sr. was a nuclear physicist who worked at the Naval Research Lab on linear and particle accelerators. He was also just as precise as his parents: Terry Jr. remembers a cache of reel-to-reel tapes made by his father that were labeled second by second.

Margaret was always available, always there, said Karen. She crocheted blankets for her great-grandchildren; she told them stories about the Delaware River flooding; she pushed Terry Jr. around in a stroller. Margaret was generous: In addition to funding an ISCC award named after I.H. for achievements in color science, she contributed to the Easton hospital where I.H. had been treated and gave annually to cancer research in honor of Esther Hurlbut Godlove, I.H.'s first wife. "Margaret, to me, had a wonderful sense of humor," said Karen. "She often had a twinkle in her eye, a sly grin, and a laugh." She mentions the definition of "begonia" from the *Third,* which I had shared with both of them when I first wrote:

> a deep pink that is bluer, lighter, and stronger than average coral (see CORAL 3b), bluer than fiesta, and bluer and stronger than sweet william — called also *gaiety*

Karen noted the whimsy of the definition, the sly wink of it. "I never met I.H.," she said, "but I saw Margaret in that definition."

The End of the Rainbow

Margaret's work as a color definer was done, but not all the color-related terms were defined. Oakes hardly had time to mark the end of Margaret's gargantuan feat of defining: There were still a handful of color names yet to gussy up. Oakes asked Dorothy Nickerson to handle the basic color names, like "blue" and "pink," as well as the rest of the colorimetry terms like "radiant energy," the article at "color," and the color plates. Margaret had suggested her or Deane Judd as possible replacements for I.H.; surely Nickerson would know exactly what the color man was thinking. At least he sure hoped someone would know what he was thinking, because what he had in hand was unusable.

I.H.'s definition of "blue," turned in in May 1954, was extensive and not particularly colorimetric. The first definition reads, "color sensation, the hue of which is that of the clear sky, or that of the portion of the color spectrum lying between green and violet (reddish blue). More specif., any of the colors normally seen when the portion of the physical spectrum of wavelengths 465 to 482 millimicrons, most characteristically 476 millimicrons, is used as a stimulus," which attempts to make blue all things to all people: a sensation, a picture, a wavelength, a stimulus—why not throw in smell and taste while we're at it? The next two definitions deal with primary colors; definition 2 notes that blue is one of *four* psychological primary hues, and definition 3 notes that it is one of *six* psychological primary hues, and neither of them talks about the three primary colors that we're mostly familiar with (red, blue, yellow). Definition 5 reads, "an object of blue color or belonging to a group whose characteristic color is blue, as a kind of Nanking china, a badge of ribbon, etc.," which is not only too broad but circular to boot: A blue thing is something that's blue.

That entry is utterly ridiculous—that is the technical term—and it's no wonder that Oakes just filed it away and decided to ask I.H. about it later. And if Oakes needed any further proof that it was Margaret who had been doing the bulk of the defining all along, here it was: an entry that came in at the very end of I.H.'s defining stint that completely ignored one of the first style decisions made for the *Third,* which Margaret had seamlessly adapted to but her husband evidently had not. I.H. ended every definition with a period, which was verboten, period, full stop.

Nickerson enlisted Judd to help with the defining, at least—Judd had already come up with working definitions for many of the colorimetric terms—and their revision was certainly more scientific. They struck sense 4 and sense 5 entirely, and revised senses 1, 2, and 3:

> **1** any color whose hue of which is that of the clear sky, or that of the portion of the color spectrum lying between green and violet which is evoked in the average normal observer under normal conditions by radiant energy of the wavelength 475 millimicrons
>
> **2** the one of the four psychologically primary hues (red, yellow, green, and blue)
>
> **3** one of the six psychological primary object colors (see primary, noun, 3)[1]

The problem with this entry was that it set the pattern for, at the very least, "red," "green," and "yellow" (which were the other three-of-four psychological primaries, a term that was so opaque no light passed through it at all), in addition to "white" and "black" (the psychological primary object colors—another term so dense it could shield the dictionary reader from radiation, and a term Nickerson apparently invented, since it's used nowhere else in any of the literature), and the remaining spectral colors, "orange" and "violet."* "Brown," "pink," and "olive," however, could be defined however Nickerson and Judd wanted to define them, and they did. All three definitions resorted to the model used for the *Second,* mentioning saturation and lightness—the model that Oakes knew didn't work and that even I.H. had abandoned.

Oakes didn't have the time or the energy to argue it. His notes on the entries—and his communications with Nickerson—were meager. Most of the buffs were just stamped by Oakes and filed in alphabetical order for inclusion. He sent notes thanking her for her careful work, and asking her to thank Dr. Judd as well, but there was no spirited back-and-forth as there was with I.H. He was being stretched thinner and thinner, and so there was less of him to give to his correspondents.

It was readily apparent, by the middle of 1955, that the *Third* was way behind schedule, and Gove was taskmastering everyone more than usual. Gove had assumed that the defining would be done by January 1956, yet in June 1955 he notes that only 54 percent of the estimated entries have been done.[2] He knew prior to 1955 that

* Nickerson also handled "purple," but remember, that's not a spectral color. She notes this in her definition, the entirety of which is "a nonspectral color." That's all you need to know about purple. A later editor inserted another definition before hers that reads, "any of various colors that in hue fall about midway between red and blue; *also* : the hue of such a color." Oakes would have had a coronary.

things were plodding and not galloping; his answer was to add more outside consultants to handle more of the vocabulary. He insists in his June report that most of them will be done within the next six months. This is arrant nonsense, and anyone who isn't Gove knows it.

Nonetheless, Gove pressed on. In 1955, he hired eight more editors, including H. Bosley Woolf: Every new hire needed to be trained, which was left to the already scraped-thin associate editorial staff. Gove grew snippier and snippier in his responses to in-house staff questions about the *Third*—they had the defining memos, for God's sake!—and he was dealing more and more with nervousness from the board regarding the reports coming in from former Merriam fans who found the new Random House *American College Dictionary* to be a much more serviceable book. The nervousness from the board oozed down into the sales staff, who began to behave in ways that some of the potential customers found ruthless or downright offensive—nasty comments made about competitors and the sensibility of a customer who would settle for something that terrible—and these complaints made their way to Gove, too. He took these complaints personally, and wrote on several occasions to professors or deans, expressing austere and slightly derisive befuddlement that these men would throw away Merriam's rigorously ordered representation of the language because some egghead in an ivory tower somewhere got it into his rarefied head that maybe the order of the listed definitions should be changed so that the definitions are given in order of most frequently used to least frequently used, rather than from earliest to appear in print to latest.* And Gove was still mired in the

* Until very, very recently, this was the Merriam way, though no one who ever consulted any dictionary I worked on in my career had any idea that the first definition was the chronologically earliest. That's been changing some in the online dictionaries—albeit, like all things that happen in lexicography, so creepingly slowly that it makes glaciers jealous.

people managing—of all his duties, the one he liked the least, and the one he felt was the biggest waste of his (very limited) editorial time.

The strain was too much. In 1956, sixteen editors resigned, including Edward Oakes.

"Dear Friends," Oakes's resignation letter begins, "I find that my disabilities have so increased that I am not able to concentrate sufficiently on the work and must ask to be retired without delay."[3] His scoliosis had worsened, and his mental fatigue and exhaustion were exacerbated by the terrifying demands of the increased schedule. Oakes had warned Holt and Bethel ten years earlier that he wasn't going to be able to keep up the pace then; his workload now was much, much heavier. He knew, however, that he wasn't dealing with Bethel but with the ever vigilant, never trusting Gove: In support of his decision, he offers to let Gove talk to his doctor to get a fuller picture. He ends his short letter by thanking his colleagues and friends, whom he calls "indulgent" in dealing with him through the years. All he asks is that he be allowed to lay his grease pencil and citation cards down and slip out of Merriam without any fanfare. "I could not bear any leavetaking," he writes, then wryly adds, "and it's not quite time for an obit."

Gove had annotated Oakes's resignation at the top with "rec'd 8 a.m. March 6 in envelope marked 'PGB' handed to me by Grace Kellogg." It is overly meticulous, even for Gove; perhaps he understood in that moment just what he was about to lose, and noting where he was and what he was doing when he received the bad news was the only way he could mark it. He then included his notes from his call with Oakes's doctor below—"by telephone (10:45)." The doctor confirmed that Oakes was in horrible and constant pain, that the brace he wore for his scoliosis was no longer helping his "badly mangled spine," and that the fatigue from both these things was so extreme that he could not carry on any longer. "But," Gove writes, "this does not mean that his brain is affected."

That is the pivot upon which the conversation turns. Gove evidently asked if Oakes might be malingering: His notes on the conversation indicate that Oakes's doctor doesn't think Oakes is exaggerating his pain simply because he wants to quit. Oakes's doctor notes that Oakes was ashamed of his low output at work: "Conscience-stricken" is what Gove has written down. Gove also must have asked if he thought that a period of rest might change Oakes's mind: "Does not believe any different decision will come after a month's rest 'but I don't know.' " That's all that Gove needs to hear.

His response to Oakes doesn't start with the usual well wishes or expressions of heartfelt concern—for Oakes, at least. He gets right down to what's most important: "Your note of March 5 is very distressing."[4] Gove goes on to say that he had indeed spoken to Oakes's doctor and had consulted with Gallan, and the two Merriam men think that Oakes should take the rest of the month off, consider it a vacation, and reconsider his resignation. Gove then offers Oakes what must have seemed like an enormous concession: come back and work until June 29, one year short of Oakes's retirement age, and then work half time from home until July 1957 so the company could continue to pay him, and so he would be better off pension-wise.

This was not a concession to Oakes, however: It was blatantly ignoring his stated reasons for resigning. He's not worried about the pension, he says: Even with leaving a year early and thereby forfeiting his full pension, he should be getting close to $1,000 a month—a figure that Gove fact-checks and notes on the envelope Oakes sent his resignation in. He lists off the reasons, again, why he cannot continue: the mental fatigue and lack of concentration, the pain. And another thing was the increased pace at work. The first two months of 1956, he explained, he had put in substantial unpaid overtime—his "best licks"—to try to meet the editing goals set out by Gove, but "even neglecting certain essential steps I never on any date reached the minimum and wouldn't if I were able to return." He also knows more about the editorial operations than Gove does: He pooh-poohs Gove's idea that the production work could be done from home by tarring the suggestion as "impractical." He is unmoved. He is resigning. "My salary," he assures Gove, "ought to get you a good editor."[5]

Gallan stepped in. He allowed him to be counted as part of the staff until 1957, though Oakes remained on leave from March 1956 until his formal retirement the following summer. Gallan's letter of thanks on the date of Oakes's official retirement was a short and honest summary of his career. "Your mark has been left not only upon the books that you helped to edit; it remains as well upon the many persons with whom you came in contact during your editorial career, especially those whom you indoctrinated in the art of defining," Gallan wrote.[6]

If Gove sent Oakes any further correspondence, he didn't keep a copy in his files. In 1956, the color work was passed off to other editors.

One of the problems inherent in dictionary work is the project timeline. Dictionary production schedules prove that time is an illusion, because they go far too quickly and stretch on forever at the same time. There is always more work to be done than can fit into a regular forty-hour workweek, no matter how many people you hire: Language is always changing, evolving, surfacing. The number of editors who saw the *Third* through from the very beginning to the bitter end was a fraction of the number of editors who worked on the *Third* in part. Of the 169 editors listed on the staff page for the *Third,* only 26 of them were there before Gove took over, and most of them stayed for only a couple of years. The *Third* took so long to produce, and so many people came and went during its production, that the staff page doesn't even list all the editors, proofreaders, and typists who ended up working on the Big Book. They exist only as pale purple date stamps sprinkled throughout the file.

The *Third* was a project-management albatross: too big, too weighty, something that stuck around for too long and reminded everyone of the futility of human endeavor. When you have that much stress, that much turnover, something will fall through the cracks.

Sometimes that something will make or break an entire subject area.

By the time Oakes had left Merriam-Webster, much of the first-round editing had been done on the color definitions, but they hadn't yet been incorporated into the rest of the dictionary entries, and all dictionary entries go through several rounds of editing. Ideally, at any rate. But these weren't ideal times. Editors kept leaving, the board continued fretting, and money was sluicing down the drain labeled "Unabridged." Time was of the essence. These successive rounds of editing, uninformed as they were, would end up smearing the canvas.

The *Third* was a house made of index cards, of which there were millions. Definitions from the *Second* were usually pasted onto buffs for editing, and these buffs were usually marked by hand. If a revised definition was big enough that it had to be written on a separate buff, then the old definition from the *Second* that was being revised was supposed to be marked with "repl": "replace."

The new buff would then be pasted up onto sheets of paper that the compositor would use when typesetting the entries.

A fine system, provided no errant breezes or editorial sighs that might topple the whole intricate structure. Any entry that has undergone multiple revisions might have multiple buffs in the file marked "repl," so an editor will look for the one that isn't marked "repl" and take that as the last and final revision in the series—sometimes without checking the date on it, or without seeing if there happened to be another buff in the file that wasn't marked "repl," too. Sometimes slips were misfiled; sometimes they were accidentally marked "repl." Sometimes hurried or busy editors made changes to slips without realizing what, exactly, they were doing; sometimes editors quit and left stacks of slips on their desk, unmarked, unfinished—unknown to anyone else on the floor. When important editors left, decades of background information held at the back of their minds shuffled out the door with them.

This was the fate of the second round of color work for the *Third*. There was the correspondence between I.H. and Oakes, if the faithful remnant still at Merriam had the time to sift through hundreds of pages of I.H.'s shilly-shallying (and provided the correspondence had been filed in such a way that it'd be easy to find). Oakes's "sacrosanct" warning memo wasn't particularly helpful, either. It mentioned which outside authorities I.H. was basing his entries on but didn't explain the highly unusual defining paradigm at all. It allowed general editors to handle literary uses of color terms, like "cerulean" to mean "a bluestocking," but this line isn't always clear—who gets to handle Homer's "wine-dark seas"? It warned that "color" was not to be used as a genus term in any definition, so no "a color, moderate green in hue," but that "color" could be used in instances where the color name was applied to a range of colors, such as "barn red" ("a variable color averaging a moderate reddish brown that is stronger and slightly redder and lighter than mahogany, yellower and stronger than roan, and stronger and slightly redder than oxblood"). It was a memo written primarily to keep people from getting their inky mitts all over the color definitions. It was not a succession plan.

The color definitions, both for names and for colorimetric terms, finally entered the second copyediting and incorporation phase in late 1955 and were overseen by the science editors Hubert Kelsey and Donald Lee, who nibbled away at the definitions like hangry carp. Why isn't "auburn" defined in terms of hair color? We've got a slip from the *Second* for "carrot orange" but only a slip for "carrot red" for the *Third;* is this a mistake? The galley says that there's an insertion No. 7 for "cracker" and we can't find the slip; is this supposed to be a color term, or a cooking term, or a chemistry term, or or or? Another editor, Walter Bezanson, was editing the Nickerson colorimetry terms with decidedly unscientific abandon. "Achromatic" gets three revisions, none of which are referred back to Nickerson or Judd—and why would they be? Nickerson and Judd had company-issue date stamps, and so many people had passed through the office since Bezanson was hired in 1954 that he assumed these were editors who did their time and then got the hell outta Dodge (as he himself would do by the end of 1956).

Because of the in-house turnover, the increasingly hair-on-fire pace Gove was setting for production, and the pesky problem of index cards going walkabout never to return, all the careful work done by Oakes and both Godloves to shape the color definitions—and therefore the primary means by which I.H. had hoped to disseminate the good news of the ISCC-NBS into the broader dictionary-buying public—was undone. Many dozens of color names that were marked for deletion by Godlove, such as "India red," were reinstated. Hundreds of color names that had been revised were misplaced, and the initial definitions that Oakes felt weren't specific enough were restored, like "melon pink" ("a vivid yellowish pink") and "flesh" ("a pale orange yellow to yellowish gray"). Some color names that had been carefully separated, like "almond" and "almond brown," were kludged together into one seemingly fine yet completely incorrect compound definition that would have inspired a twelve-page-long rant from I.H. about the differing industrial and textile uses of "almond" and "almond brown" had he still been in any condition to rant. None of the editors, it seemed, realized that there was a Godlove still living who would have been happy to answer any questions, and why would they? She was, after all, a mere stenographer in the eyes of everyone else on the editorial floor.

The last color definition to be edited was, appropriately enough, the short entry for "color," which was turned in in 1957. There were two main definitions Nickerson and Judd supplied to Merriam-Webster, a summation of the last hundred years of chromatic learning, experimentation, and labor in nine lines. One covered the uses of "color" that had to do with how our brains produce color:

> **2** *psychology* **:** an aspect of the appearance of objects and light sources described and specified in terms derivable wholly from one's perception, most commonly involving hue, lightness, and saturation for objects and hue, brightness, and saturation for light sources[7]

This definition was only lightly edited; the label "psychology" was removed and was turned into a usage note at the end: "used in this sense as the psychological basis for definitions of color in this dictionary." "Commonly" is also weirdly changed to "conveniently," as if the appearance of the object or light source were somehow in cahoots with your perception, and it's *most convenient* that this all coalesces into something we call color. But if you had to ask a scientist to give you a definition of color to match the word in the sentence "The color of that ball is bright blue," you could do (and had done) worse.

The second definition they sent in was slightly more convoluted, and was for the sense of "color" that has to do with the actual way that color, as science understood it, worked:

> **3** *psychophysics* **:** the characteristics of light apart from variations in space and time, light being the aspect of radiant energy of which the human observer is aware through the visual sensations that arise from the stimulation of the retina of the eye **:** the psychophysical correlate of the psychological term color (see color 2) identified, for spectral colors, by dominant wavelength, luminance, and purity, and for nonspectral colors (purples from red to violet) by complementary wavelength, luminance, and purity

This is a highly complicated, barely readable definition, but it is nonetheless a tidy distillation of the main thing that all of color science had learned about color in the last century: that color is an interaction between light, the eye, and the brain; that the product of this interaction can be objectively measured; that there are spectral colors and nonspectral colors (and what the nonspectral colors are). Once you sit with it, it's actually a marvel of science communication.

It is heavily revised before making it into print. What appears in the finished copy for the *Third* is this:

> **3 :** the characteristic of light by means of which two areas of identical size and shape that are juxtaposed, structure-free, and steadily and uniformly illuminated may be distinguished by a human observer and which is commonly identified for spectral colors by dominant wavelength, luminance, and purity and for nonspectral colors (as purples) by complementary wavelength, luminance, and

> purity—used in this sense as the psychophysical basis for measuring color which in turn makes it possible to define the limits for each color definition used in this dictionary; see the Color Charts[8]

There is no record at all of why this revision is needed; it appears ex nihilo, and, like the unformed darkness whence it comes, it's confusing and disorienting as hell. What are "areas of identical size and shape," and how are they "structure-free"? Can structure-free areas (whatever those are) be juxtaposed? Most important, how is the "human observer" distinguishing this characteristic of light? The point of this definition is to mark the whole psychophysical process from light to eye to color stimulation. Were I a jerk (and I am), I'd argue that, according to this definition, I could perceive the color of two identical balls by having a color technician shout their dominant wavelengths, luminance, and purity to me, and that this is a way to "distinguish" color. The usage note, too, is a strikable mess. "Define the limits for each color definition"? This is amateur hour. Who allowed this hideosity?

This revision was made by Gove himself in 1958. So much for "sacrosanct."

The only chance to salvage this mess was the color chart. As in the *Second,* the *Third* allotted two color plates for the color definitions. A good two-page chart, with color chips for the reference colors the Godloves used as data points for every color-name definition in the *Third,* would have at least made the Godlovian system clear. Anne Driscoll was given the charts, and she handed them to Dorothy Nickerson, who knew nothing of the Godloves' plans for the charts; indeed, it's likely that neither I.H. nor Margaret had had time to make concrete plans for the charts. She did her best. There's the illustration of a full-color Munsell solid from two sides (an illustration she got from Munsell by calling in a favor); there's one wedge of the Munsell solid divided up into the ISCC-NBS notations and some sample charts for that notation overlaid on a color; there's a copy of the visible spectrum, repainted by Charles Bittinger; an example of the Munsell hue versus value axes; a Munsell hue circle. The explanation is found in the essay she's submitted for the *Third,* which will snug up against the color plate on the verso; a feat of editing once you realize that she managed to condense the entire ISCC-NBS *Method of Designating Colors* down into twenty-seven column inches and change. But color chips for "holly green" or "cerise" or "aloma"? Nowhere to be seen.

Thirteen years after I.H. first wrote back to Merriam-Webster and offered to write definitions for the *Third,* the color definitions had been revised according to the latest, best science, yet bent in the direction of common commercial use (in accordance with Govian principles); the color terms written by three of the most distinguished colors scientists of the twentieth century; and the color plates created in accordance with the standard adopted by the twenty-one member societies of the ISCC, the NBS, and the American Standards Association. Every part of this subject had the scientific stamp of approval on it. And thanks to the vagaries of production schedules, the disruptions of staff resignations, the pressures of budgets, and the cussedness of Gove, thirteen years of careful construction were knocked askew.

A rainbow of more than three thousand colors, lovingly delineated, unwittingly smeared into gray.

The Color of Money

The *Third* was finally released in 1961: two years late, a million dollars over budget, scads of editors crushed under its heavy, churning wheels. The only place where the dictionary came in under estimate was its length. When Gove took the helm, he had 3,400 pages to work with. The finished product was a lean 2,716 pages.

After Oakes's retirement, the staffing picture only got worse and worse. Editors left almost as soon as they were hired; in the last six months of 1959, while the editorial department was pushing and pushing to get pages to the typesetter, eighteen editors jumped ship. Gove calculated that this was turnover of more than 30 percent. In a 1959 report to the board, the situation is so dire that Gove decided to break ranks and speak out of turn: "Many staff members have been pushed to the limit of their capacities, and even some beyond. They are tired. I set this down as a fact with no analysis of its possible significance to our work." It is as empathetic as Gove got.

And yet, in the copy of that June 1959 report that Gove retained, he starred the paragraph with that statement in it, and sent it along to Gallan first, with a note scrawled at the bottom. "This [paragraph] can, of course, be omitted, but it is, I think, very much to the point."[1] It appears that Gallan left it in.

The publication of a dictionary, and especially a pricey doorstop of a dictionary, is generally heralded by few—mostly just the people involved in making said doorstop. At best, most dictionary releases get a day or two of soft-focus back-patting in the press, and then they fade into the fusty obscurity of trade journal reviews. A little copying from the press release about how many great new words are in, and a (carefully curated) sample of those words, and then it's back to other, more proximate news stories: the scandal of Mrs. Grundle's new lawn ornaments, for instance, or the tedium of the school board's latest budget.

And so it likely would have gone for *Webster's Third,* had Merriam's president, Gordon Gallan, not decided he needed a big marketing push to recoup some of the costs sunk into his doorstop. He hired a public relations consultant to write (with an assist from the in-house marketing team) a zippy press release highlighting all The New inside the *Third*. Gove supplied some basic facts about the new, scientific, fully objective treatment of the *A–Z,* while Gallan fed the consultant some impressive-sounding numbers to sprinkle throughout: 2,716 pages, 450,000 total entries, 13.5 pounds, costing $3.5 million to produce, more than 200 outside consultants, and more than 757 editorial work years to complete. He was banking on the press release making a splash, and it certainly did. It cannonballed.

The problem with the press release for the *Third* was the problem with all previous press releases for all unabridged dictionaries since the early nineteenth century: It was a press release. The goal is to drum up interest, and subsequently sales, in a book that most people, over the course of their entire lives, bought exactly one of. Everything that the *Third* did was called "revolutionary," a word that appeared multiple times in one press release: author quotations (not new to Merriam nor, to anyone who happened to glance at the *Oxford English Dictionary,* revolutionary), the new defining style (admittedly new to Merriam, but was it really going to usher in a revolution?), the pronunciation guide (simplified but, as letters to the company would later make clear, still too damned hard to decipher, and not exactly a pronunciation guide that was going to change the world). Gove's changes were entirely oriented toward high scholarship, but the press release billed the dictionary as a book that was "Planned to Be Read and Enjoyed by Students, Housewives, and Business Men."* And if the marketing folks thought that this was a book to be enjoyed around the dinner table, they certainly weren't going to highlight entries like "colorimetry." Instead, they excerpted part of the definition of "ain't" and highlighted other slang entries as a fun way to connect with the common man.

It certainly connected—like a right hook from Muhammad Ali. The press mocked the slangy terms that appeared in the prepublication press release. "A passel of double-domes at the G. & C. Merriam Company joint in Springfield, Mass., have been confabbing and yakking for twenty-seven years—which is not intended to infer that they have not been doing plenty work," began an October 1961 editorial in *The New York Times* that razzed the dictionary for not judging language quite enough for the public's taste by highlighting some of the new(ish) terms that appeared in the *Third*.[†2] It set the topic and the tone for nearly all other reviews, pre- and

* Having worked on the dictionary in question, I can assure you it is no such thing unless your family or supervisor is particularly cruel.

† Gove responded in a letter to the editor, because of course he did. "The paragraph is, of course, a monstrosity, totally removed from possible occurrence in connection with any genuine attempt to use words in normally expected context. It hits no mark at all." Sounds as if it hit *some* mark, Dr. Gove.

post-publication, of the *Third* that were seen in the general press. Newspaper reviews ignored the leap in linguistic scholarship that Gove had put together and instead focused on the informal words or less-than-elevated authors (like Mickey Spillane and the infamous madam Polly Adler) quoted in the Big Book. "Ain't" and the apparent open-armed embrace of it in the *Third** was the most notable flash point around which all early coverage exploded, but the informal words were highlighted in the press for years: EXPERT SAYS 'NECKING' GOOD WORD, reads one headline about the *Third* from 1964.[3] At least the headline writer called the editors experts.

Gove was a man who did not give an inch, nor suffer fools, so he did not take these boos lightly. How could he? Here he was, his life's work complete, his desk empty of all things W3, the office increasingly desolate of people and permeated by the scorched-coffee fug of burnout. It didn't matter that the newest edition of the *Collegiate Dictionary* was in final read-through; it didn't matter that, as he had bellowed at editors and then reporters, the language keeps moving and so does Merriam. The *Third* was his book; it was, in a very real way, *him*. And so we can perhaps understand why he spent the next several years writing detailed letters to the editor in defense of his Big Book (and himself). When the press slung slang or deplored that The Dictionary had given gravitas to that business jargon bastardization "finalize," Gove sneered that "the new edition merely reflects the tremendous change that takes place in a language when 180 million fairly literate people are speaking it."[4] Gove's written responses to bad reviews of the *Third*, fizzing with the icy combination of erudition and irritation well known by his editorial staff, appeared notably in *The New York Times* and *Life* magazine, but he continued to type up and send out responses to

* This misapprehension rests at the feet of the really, truly terrible press release. If you want a play-by-play, Herbert Morton's *Story of Webster's Third* and David Skinner's *Story of Ain't* are the books for you.

smaller papers whenever a bad review crossed his path. It did not win him or the *Third* many fans.

But the *Third* had its strengths. Linguists especially noted, mostly in academic journals that no one read, that it was constructed on sound and modern principles, and no one could miss how much technical and scientific vocabulary it covered. "Wars, fashions, the penetration of space and the astounding advances in science, industry, medicine and other vital aspects of human existence have forced new words upon us," one syndicated prepublication review of the *Third* noted, and that the dictionary had kept pace with it all was a marvel.[5] The journal *Science* printed a review of the book that essentially rehashed much of the popular criticism of the dictionary but allowed that the *Third* had paid its "debt to science more fully than general culture."[6] Even the purveyors of general culture agreed: Despite the sneery pan published in 1961, *The New York Times* noted just a few months later (in another pan),

> Editors representing the news, Sunday and editorial departments have decided without dissent to continue to follow Webster's *Second* edition for spelling and usage. Webster's *Third* will be the authority only for new, principally scientific words.[7]

But the whiz-bang technical vocabulary didn't exactly win over the literati, who saw something more insidious in the *Third*. It began with Wilson Follett, a grammarian and usage specialist who savaged the dictionary in his review for *The Atlantic* in January 1962. The review was headlined SABOTAGE IN SPRINGFIELD, and it was not hyperbole. Sure, the coverage of scientific terms was great, Follett noted, but at what cost? Follett saw Gove's stated linguistic approach—to describe the language as it was used rather than offer advice to the reader about how to use it—as an abrogation of authority. It is the duty of the lexicographer to make sure that the record of the English language doesn't get sullied with crapola used willy-nilly by the hoi polloi.* "With but slight exaggeration we can say that if an expression can be shown to have been used in print by some jaded reporter, some candidate for office or his speech writer, some potboiling minor novelist, it is well enough credentialed for the full blessing of the new lexicography," Follett frets. It's not a mistake that he goes from scientific and technical language right into his big complaint about Dictionaries These Days; at the end of the piece, he writes that the lexicographer can "think of himself as a detached scientist reporting the facts of language," but no one else thinks of the lexicographer this way. They don't want the new science; they want guidance.[8]

It started something of a dictionary-based moral panic. Dwight Macdonald, social critic and writer for *The New Yorker,* came out with a scathing review of the *Third* and what it was doing to American culture in 1962. His essay "The String Untuned" is famous in lexicographical circles for reviewing the *Third* not just as a dictionary, but as an example of what happens to culture when science grabs hold of it. "This scientific revolution has meshed gears with a trend toward permissiveness, in the name of democracy, that is debasing our language by rendering it less precise and thus less effective as literature and less efficient as communication," he writes.[9]

Later that year, Jacques Barzun, a well-respected cultural historian, derided what he called the "scientism" of the *Third,* taking, as an example of it, the use of the swung dash (~) in example sentences in place of the base form of the word being illustrated. This

* Follett, if he were alive today, would see that definite article before "hoi polloi," lick his pencil tip, and write another long screed about how the lexicographers who came out of the Merriam shop are continuing to machete the language to bits. "Hoi polloi," as any good pedant will tell you, already has a built-in article in Greek (the "hoi"); hence, "crapola used willy-nilly by hoi polloi." But Follett is dead, and I am an unrepentant producer of crapola.

is, as anyone at Merriam will tell you, a space-saving convention: You can gain a lot of characters if, instead of printing the example "underwent electroencephalographical testing" at the entry for "electroencephalographical," you use "underwent ~ testing" instead. Incorrect, proclaims Barzun. Taking the entry for "of" as an example, with its many swung dashes in the example sentences, he explains that this typographical convention is "a subtle attack on The Word, just as it is a blow struck against the sentence. The dictionary ceases to be a book for readers interested in words and sentences; it becomes an imitation ~ the technical handbooks ~ physics and chemistry. By this means words . . . are reduced to algebraic signs."[10] He does not think this is a good thing.

These criticisms weren't just based on the personal peevery of the authors in question (though plenty of that shows up in each review). There had been a growing unease with science, and our increasing reliance on science, since the end of World War II. In the 1940s and 1950s, there were some who saw the supposed objective and observational position of science as value neutral in places where neutrality was not kosher: patriotism, for instance, or anti-communism, or morality. Some things, critics claimed, did not need the cold eye of objectivity staring them down. As McCarthyism swept through the American landscape, this sort of objective distance was seen by some as evidence of communist sympathies; left-leaning scientists and social scientists, in particular, were suspect.

Into this wariness came the field of structural linguistics, where language—that ineffable muse of Shakespeare, Dryden, Pope, that sunlit heath upon which Dickens and Twain frolicked—could be reduced to a carefully constructed scaffold upon which words were hung like wet laundry. The father of modern semantics claimed that this scientific evaluation of language would mean that the general language user could "abolish his old-fashioned 'private opinions'" about what language was and wasn't, or what language was good or bad, when linguistics could handle that scientifically.[11] Some folks did take things a little far, to be fair: C. K. Ogden and I. A. Richards, late of the "triangle of meaning," put their heads together and proposed shrinking the English language down to 850 words. That would be plenty, they reasoned, to describe movement through space, and what more did you need for communication?[12] If there was a furor from the humanities folks, well, they'd get over it, surely.

But they did not get over it. By the time the *Third* came out, there was a flourishing revival of the humanities in American intellectual life, and an accompanying criticism of the cultural impact of science. Specifically, the humanities folks claimed that science was destroying the very thing that made us human—the ability to feel, to create art, to think outside the box, to communicate—by quantifying it, gatekeeping it, and irreparably changing it. Language, as a tool of communication, got a lot of attention.

Barzun's 1962 attack on the *Third* was just a warm-up. In 1964, he released a book called *Science: The Glorious Entertainment* in which he skewered scientism in language by coming back, once more, to the *Third:*

> The dictionary is no longer a list of words spoken by the people, but a semi-encyclopedia, in which a score of technical vocabularies lead an autonomous existence, together with initials, abbreviations, and trade names. In the holes and corners sulks the native tongue, heavily laced with slang, the jargon of war, and popular misconceptions of scientific terms. . . . Whether [the language user] knows it or not, language is the keeper and shaper of his consciousness, and he cannot habitually cast his will and fears and hopes in broken or foolish patterns of words without becoming a foolish and discontinuous mind.[13]

So if science was degrading our ability to communicate, and the dictionary abandoned its (humanities-tilted) post as the record of the best practices of the language in favor of, as Barzun overwrites, "technical vocabularies [that] lead an autonomous existence," then of course you get entries like this in a dictionary that gives its crown to science:

> a nonmetallic chiefly bivalent element that is normally a colorless odorless tasteless nonflammable diatomic gas slightly soluble in water, that is the most abundant of the elements on earth occurring uncombined in air to the extent of about 21 percent by volume and combined in water, in most common rocks and minerals (as oxides, silicates, carbonates), and in a great variety of organic compounds (as alcohols, acid, fats, carbohydrates, proteins), that has three naturally occurring nonradioactive isotopes of masses 16, 17, and 18 of relative abundance 2494:1:5, that is obtained industrially from liquid air by distilling off the nitrogen or from water by electrolysis or in the laboratory by decomposition by heat of various, oxides, peroxides, or salts (as chlorates or permanganates), that combines with all other elements except those of the group of inert gases, and that is used chiefly in oxyacetylene and oxyhydrogen flames in welding and cutting metals, in making steel and in other metallurgical processes, in making glass, in the chemical industry (as in producing synthesis gas), in medicine, aviation, and diving to aid respiration, and usually in the form of air in many combustion and oxidation processes — symbol *O*

That monster is, in case you aren't sure, the definition for the element "oxygen" as it appears in the *Third,* and it is almost as long and breathless as the definition for "oxygen" in the *Second*. Where Barzun saw the cold, mechanical hand of science, the discontinuous minds at Merriam (and probably Gove himself) saw one of the biggest weaknesses of the *Third:* Given the staff turnover, the overreliance on outside consultants, the vastly changed defining style, and the lack of institutional knowledge, a whole lot of encyclopedic information still snuck into Gove's dictionary of "pure language," and entries like "oxygen" were a prime example of it.

But a handful of people noticed: Laurance Hart, for instance, who called himself the "Lexiconoclast" and reviewed dictionaries, atlases, and other reference works. In his mimeographed review of the *Third,* he notes that much of the encyclopedic matter has been axed, but "much of what remains is dubious." He mentions, as a primary example of this, the scientifically extensive and absolutely un-Govian entry for "color."[14] Good thing Barzun didn't make it to the letter *C.*

And this was, speaking of the letter *C,* the real catch-22 of the *Third*. Gove wanted a dictionary written according to the best linguistic and scientific principles available to usher in the new age. He crafted a defining style so abstracted, a method so rigorous, and a timeline so impossible that few of his in-house editors had the time to correctly execute Gove's vision. In a time crunch as inescapable as a black hole, the staff had to let the outside consultants do whatever they were going to do so that they could instead rework every general vocabulary entry in the dictionary so that it met Gove's exacting standards. These outside consultants were themselves children (or fathers) of the new scientific age and had no problem taking the space afforded by an unabridged dictionary to represent their fields in full. Gove, so enmeshed in the minutiae of running an editorial floor and hating every minute of it, didn't give his book the oversight that it really needed; Gove, so enmeshed in his own rightness, didn't think that anyone's criticisms of the Big Book were valid, because those people were not Gove. The *Third was* full of technical vocabulary, but it wasn't the result of a dastardly plan to make communist robots of us all. It was the unfortunate consequence of one man's nearly monomaniacal desire to get language right, pure, and clear—all language.

Gallan, who was charged with selling this damned book, thought all this was a double-edged sword. On the one hand, any press is good press, and people *had* gone out to buy the Dictionary with No Standards. On the other hand, the core purchasers of reference books—colleges, newspapers, libraries, the government—were not exactly keen to purchase a book that was seen by the general populace as essentially a gimmick. The folks who appreciated Macdonald and Barzun were exactly the types of people who would buy, use, and eagerly promote a dictionary they loved. All this gobbling over the scientism of the *Third* stung.

So perhaps he was done with scientists when Dorothy Nickerson wrote to him in March 1962 to complain about her treatment at the hands of Merriam-Webster. At the end of 1961, in keeping their promise to Nickerson, Gallan had copies of the color plates sent to her. Of the dictionary itself, she says nothing ill: She has already referred plenty of people to it for its superior treatment of color terminology, and she's happy with how the color charts came out. But "it was something of a shock to me, following our correspondence of 1958 to receive only six copies of these prints [of the color plates]. . . . My letter of June 24, 1958 talked in terms of an extra few thousand, certainly not an extra half dozen!"[15]

She tried to be generous—they probably didn't have many overruns to spare—but she agreed to take on the task of creating the color charts only if she could use them for her own purposes afterward. Nickerson had always wanted to further color education; these charts, set up as handy one-page intros to the Munsell system, the idea of a color space, and a shorthand for the

ISCC-NBS method, were perfect. She reminded Gallan that when she got Munsell to supply high-quality prints of the solid and other illustrations for the final version of the chart, it was at no cost to the company. "From my point of view," she finished, "you have hardly complied with the spirit of the correspondence we had on the subject."

Gallan, in full harrumph, sent along fifty extra sheets for her, blaming the manufacturing strain of printing the *Third*. But he will entertain no other complaints or requests on this topic. And to add insult to injury, Gallan reminded Nickerson that the company prohibited reproduction of any of their color plates without permission, and that all of the material on the color plates is covered by Merriam-Webster's copyright. Perhaps one of Merriam's many lawyers could have reminded them that, actually, Munsell and the NBS held the copyright to each of the images on the color plates. Nickerson had collected them herself and gained permission to use them because Merriam had promised to send her thousands of copies of the finished plates for Munsell's and the NBS's use. Merriam holds no formal licensing agreements from either Munsell or the NBS.

So you can understand why, when she was contacted by a representative of James Parton's in 1964 and asked if she might be interested in providing the color definitions for the *American Heritage Dictionary*—a dictionary that promised to right all the wrongs of *Webster's Third*—she agreed.

The gestation and creation of the *American Heritage Dictionary* have been detailed in other, better books, but the gist is this: James Parton, publisher of the *American Heritage* magazine and supporter of Macdonald's and Barzun's thoughts on the *Third,* wanted to sink the Big Book. He tried a hostile takeover of Merriam—one in which he explicitly told shareholders he intended to pulp the *Third,* reprint the *Second,* and right the ship—but failed. For most people, that would be the end of it, but not James Parton: He tapped William Morris, a well-known usage maven, and told him they were going to make their own Big Book to compete with the *Third*.

Planning for this new dictionary, which Parton called *The American Heritage Dictionary of the English Language,*[*] began in earnest in 1964. It was to be an unabridged dictionary, a return to the encyclopedic dictionaries of years past, and it was also going to use the best experts in the country for its scientific and technical language. They began hiring lexicographers from other companies (including Merriam, which had shed disaffected but well-trained lexicographers the way a sinking ship sheds rats) and started collecting consultants, including some who had worked for Merriam on the *Second* or *Third*.

Nickerson was not exactly thrilled but amenable. The plans were to give her as many color plates as she wanted. Could she define the colors and the color terms according to the ISCC-NBS standard? American Heritage thought that was grand—she was the expert.

And so she did. The major colors of the ISCC-NBS standard were entered along with a number of colorimetric terms. She kept all her defining slips and filed them away for the future. It's fascinating to compare her color definitions with the Godloves', if only to see that the ISCC-NBS standard could be interpreted in so many different ways. "Firmament blue" in the *Third* is "a pale blue to a light greenish blue"; in the *American Heritage Dictionary,* it is "light greenish blue to strong purplish blue."[16] Where'd the difference come from? The ISCC-NBS, in its very final post-Godlove form, included the colors from the Plochere Color System, which was used in interior design: The purplish blue was theirs.

* It is a reality as certain as death and taxes that lexicographers will forever be naming their dictionaries in homage to Samuel Johnson's *Dictionary of the English Language*.

Sometimes Nickerson's specificity, her desire to be accurate, was unhelpful. Her proposed definition for "eggplant" reads "dark grayish purple to dark purplish red to bluish purple," which to the layperson just looks like a lot of flailing around near "dark purple."[17] Margaret's definition, "a variable color averaging a blackish purple," is less specific, but easier to read.[18]

In 1965, Nickerson met with some of the editors to talk about the color charts—specifically, using the color chips keyed to the ISCC-NBS pocket colors, which were, she noted, sorely lacking in the Merriam dictionary. The American Heritage Editorial Board was absolutely delighted: Yes, give us everything that Merriam denied you, they practically swooned. It looked as if the ISCC-NBS standard was finally, finally going to be fully implemented into a dictionary, and thereby spread into broader use.

But in 1966, just as she was turning in all her definitions—the first consultant to do so, thank you very much—plans changed. Writing a dictionary is, it turns out, very expensive, and this came as a surprise to Parton and his board (though, given that the damned editorial costs for the *Third* were printed in the press release, it should not have been). Everything that seemed to befall Merriam during the *Third* was now befalling its main new competitor. To try to conserve costs, the budget for the dictionary was radically slashed. The company was plagued with personnel problems—not because it had a managing editor who would rather set fire to the building than allow socializing in the office, but because the work was difficult, the pay was low, and there wasn't, as there had been at Merriam, an established workflow and understanding of just how long it takes, for instance, to review the definitions from the science consultants. Consequently, the *American Heritage Dictionary* was also, like all big dictionaries, perpetually behind schedule.

A radical decision was made: In order to get this book out sometime before the universe ended in fire, the encyclopedic unabridged was going to be turned into an abridged dictionary. The marketing folks decided it would be sold as a "family-reading" dictionary (another lesson not learned from the Merriam press release); the editors decided it would be *slightly* encyclopedic, but not nearly as technical. Color definitions like "eggplant" were revised to be simpler ("a blackish purple"), while others were dropped altogether. The color plates were axed: In the end, Nickerson got one black-and-white-only column in which to share all the information about the ISCC-NBS system used in defining and no index of color names defined in the dictionary. She was sore for years afterward that "metamer," "either of two colors that appear identical to the eye but have different spectral composition," as it's defined in the *Third,* was cut.*

In her complimentary copy of the *American Heritage Dictionary* and on her defining slips, she has marked every single revision, omission, and change in green pencil, in preparation for the next edition. That careful work was for naught: The colors, having already been adequately defined for the first edition, would not be revised except to be dropped as space was needed. A color consultant was never brought back in: One of the original in-house editors, Olga Coren, handled any color revisions that needed to be made for the third edition, and as far as the last (and now former) executive editor of the *American Heritage Dictionary* knows, they haven't touched the color names since.

Nonetheless, Nickerson's work, though truncated, meant that the two most talked-about dictionaries of the twentieth century had (some) color definitions, and (most of) a paradigm for talking about color, that were written entirely according to the ISCC-NBS standard.

* Not only did she keep her defining slip for "metamer" and note it had been "omitted," but she made an annotation in her complimentary copy of the *American Heritage Dictionary* highlighting its absence.

The Gold Standard

Merriam-Webster has some sort of architectural *thing* for high places. Merriam-Webster's new building on Federal Street in Springfield was not eight stories tall, as the Myrick Building was, but it was built on top of a hill. Both offices—Carhart's aerie near the Connecticut River and Gove's corner office on Federal—give a bird's-eye view of Springfield, of the bustle below, the rolling farmland beyond the river. It's very easy, from that lofty perch, to think that the world revolves around you and your work. It's very easy to forget that there's always a taller building, a higher hill that other people have their eyes on.

Merriam-Webster was not Margaret's only color-related gig. In 1955, Margaret joined the color firm Davidson and Hemmendinger as a laboratory assistant. The Davidson and Hemmendinger in question were I.H.'s old colleagues at General Aniline Henry Hemmendinger and Hugh Davidson. The color defining was not going to pay bills, and besides she liked Henry and Hugh.

Davidson and Hemmendinger were both let go from General Aniline in 1952 as the budget for research was razed. They hung out a shingle together and began offering color measurement, color standardization, and color education services, but with some hefty scientific chops behind it. Hemmendinger's interest was in metamerism, the phenomenon where two colors appear identical but aren't (and one of Nickerson's *American Heritage*–related sore spots).* Davidson's claim to fame was the creation of the Tristimulus Computer, a computer that converted complicated spectrophotometric color measurements (used mostly in science) into a three-coordinate measurement that approximated what the eye saw. And one of the places where their interests overlapped was in gloss.

Gloss had been a big problem in colorimetry since the beginning. Anyone who has ever had to paint a room knows the difficulty of gloss. When you buy interior paint, you don't just have to pick a color from the vast Technicolor wall of samples; you need to pick a paint finish: flat, eggshell, satin, semigloss, or high gloss. You decide to paint your room Rejuvenate, a lovely sage green, and just pick a finish because you are tired of being at the hardware store and *good Lord* it was hard enough to pick a *color*. You choose "satin," take the paint home, and paint your room. The paint, when dry, appears richer or darker than the color of the chip you chose. The chip is in a flat, or matte, finish; the higher up the gloss gauge you go, the richer and darker a color appears. This goes against what you expect: With a glossier finish and more light reflecting from the surface, you think that the color should actually appear lighter or brighter, not darker. The only time that your satin-finish wall looks lighter than the paint chip in Rejuvenate that

* There are two kinds of metamerism. Illuminant metamerism is when two colors appear identical under one light source but are different under another. This is the metamerism of predawn sock-matching attempts, where you put on two black socks getting ready for work and discover, under the harsh light of the office fluorescents, that you're wearing one black sock and one blue sock. Observer metamerism is when two colors appear identical to different people. This is the metamerism of the sitcom trope where the wife claims the husband's gross recliner doesn't match the new couch and he claims it does because he's not giving up that recliner, Barbara.

you're angrily clutching in your paint-stained hand is when you sit down at your desk and turn on your desk lamp—and then you are caught off guard by the shiny reflection directly behind the lamp, which will, now that you've seen it, forever draw your attention away from your computer monitor and to this bright spot on the wall, like a moth to reflective flame.

Gloss depends ultimately on two kinds of light reflection: specular and diffuse. Specular reflection is the fancy way of talking about regular reflection, like what you get from a mirror. It has a direction: Light hits an object's surface at a particular angle and is reflected back at a complementary angle. That's the bright spot on your wall, which is absolutely not Rejuvenating the longer you look at it. Diffuse reflection, by comparison, has a million directions: The light hits the object's surface at a particular angle and is reflected back in every direction. The reflection is diffuse, scattered. When it comes to gloss, the higher the diffuse reflection and the lower the specular reflection, the flatter or more matte the surface appears.

This reflective weirdness created lots of problems for colorimetrists when they attempted to measure the color of a glossy surface. Scientists had already defined what "gloss" was and how to measure it, and created glossmeters to do just that.* But glossmeters measured only the amount of gloss on a sample; they didn't reckon with what gloss did to color.

Gloss creates a whole new layer of complication when you're trying to match a matte and glossy color to a particular color standard with a given color name. Go back to your room in Rejuvenate and your fistful of paint chips from the hardware store. Your chip in Rejuvenate, flat finish, is lighter and looks more washed out than the Rejuvenate on your wall (satin finish, remember, because

* Glossmeter, colorimeter, tintometer—perhaps the one place where scientific language is predictable is in naming instruments that measure things.

you thought it sounded luxurious and fancy). You go through paint chips, comparing them with the color on the wall; Green Tea, flat finish, just about matches. So what color is on your wall: Rejuvenate or Green Tea? The label says one thing and your eye says another.

If only someone would standardize color names across all the types of gloss finishes, you sigh, and are joined by thousands of color manufacturers, sellers, forecasters, and scientists. High-gloss coatings for use on cars were first produced in the 1950s, during a time when the auto was no longer just a means to get from one place to another but a personal style statement, but the coatings weren't very durable and after a few years would need to be regularly and aggressively waxed to stay shiny.* We were in the nuclear age, the sci-fi era, where bright colors, bold patterns, and shiny metallic highlights dominated interior design. You could get pastel fridges and teal metal cabinets and Formica countertops and sparkly vinyl seats—each one made from a different material with a different type of gloss, but all needing to match across manufacturers. The second postwar color boom had all the color-matching problems of the first one, just shinier.

The NBS, and hence the ISCC, had been trying to create a standard for glossy colors since the 1930s. A few things intervened—a depression, a war—and in the meantime other color companies began to dip their toe into the glossy-standard pond. In 1950, the board of the Munsell Color Foundation issued a policy directive that made it clear that the Munsell Color Company's chief purpose was to "develop and supply accurately controlled color standards," with an emphasis specifically on creating a glossy version of the

* And that's why, in movies of the 1950s–1980s, men are forever washing and waxing their cars.

Munsell Book of Color.[1] In 1952, the company commissioned Davidson and Hemmendinger to prepare paint formulas and produce samples for a thousand colors in both matte and glossy paints. Blanche Bellamy, the manager of the Munsell Color Foundation, reported in 1954 that both sets of chips were measured spectrophotometrically, with the idea that comparing the measurements of the matte and the glossy colors would lead to more formulations and samples that would end up providing a matte *and* a glossy standard. The work on the matte chips was completed in 1953—on to the glossy chips.

They were not daunted, because they had a secret weapon in their pocket: Margaret Godlove. Both Davidson and Hemmendinger knew Margaret well: They had sat in her den, drank her coffee, watched her manage the defining work and the Jubilee Issue, talked with her and I.H. about the latest issue of the *Journal of the Optical Society of America*. They knew I.H. had helped Granville with the measurements and calculations for the *Color Harmony Manual;*[*] they were both aware that the *Manual* included glossy chips, which I.H. would have taken spectrophotometric measurements of. They knew that Margaret had been taking measurements of color chips and, at the very least, helped I.H. calculate the Munsell notations for the ISCC-NBS colors for the Merriam defining. Margaret would have known how to operate most of the measurement equipment, been up to speed on all the difficulties of glossy standards, had the math skills to be able to see errors in the Munsell notations, and (as the ISCC newsletter secretary) had the contacts to easily find a vendor to help produce the chips for the glossy standard.

She was a natural choice: Margaret joined Davidson and Hem-

* A surprising thing to have learned, considering his mild derision of Granville's *Color Harmony Manual* in all the Merriam-Webster correspondence. He was a Munsell man to the end.

mendinger just in time for the high-gloss fun. Her 1958 alumni report to Oberlin College notes her new employer; on the line that reads "nature of occupation," she has written "Color specification (spectrophotometry)."[2]

It is the established procedure, when researching in archives, to do so quietly. This guideline holds at the Hagley Library and Archives in Wilmington, Delaware, as well. No matter that the library is located at the old site of the DuPont Company's first gunpowder mill: Explosions are unwelcome.

So let me formally apologize to the Hagley staff, and the gentleman who was, in fact, researching industrial explosions two tables away from me, for the day when I first cracked open the Henry Hemmendinger papers, flipped through the data and formulations for the *Munsell Book of Color,* glossy edition, and then slapped the tabletop hard and hollered aloud.

Margaret herself kept no records of the work she did for Davidson and Hemmendinger; in fact, she kept none of her own work, but only I.H.'s. But Henry Hemmendinger kept meticulous records of all the data and formulations compiled for the glossy *Munsell,* and those records show just how instrumental Margaret was in the creation and compilation of the glossy standard.

First, Davidson, Hemmendinger, and Margaret mapped out the Munsell colors they were going to replicate in a glossy standard. They needed their baseline. They already had the Munsell notations that Margaret used for the *Third,* and that gave them three key measurements to aim for: hue, value, and chroma. They also had the spectrophotometric measurements of the matte Munsell chips, which measured how much light is reflected back from or absorbed by the sample across every point of the visible spectrum—and a stock of the matte chips. They then took measurements of each chip with Davidson's Tristimulus Computer (and why wouldn't they?), and that gave them a measurement that approximates how we perceive that color.

That's all well and good in terms of *measuring* a color, but it doesn't help you *make* a color. Tristimulus coordinates can describe only how color is perceived rather than what the pigment makeup of that color is. It's the difference between looking at Ellsworth Kelly's work *Blue Panel* and saying, "That's a big, blue parallelogram," and looking at *Blue Panel* and saying, "That's 32.5 percent phthalo blue, 16.4 percent ultramarine blue, 46.7 percent cerulean blue, and 4.4 percent flake white."* What was needed, then, was some sort of formula to take the spectrophotometric measures and the tristimulus measurements and overlay them onto a system that would allow them to scientifically and accurately produce a color.

Fortunately, this system existed, and it even took one type of gloss into account. In 1931, the CIE published what they called the "XYZ" color space, which was a model that linked the physics of color (their wavelengths) and the psychology of color (how we perceive a color).† If you knew how a color was perceived and what the wavelength of that color was, then you had a good chance of figuring out how to make that color. If Davidson, Hemmendinger, and Margaret could figure out the CIE values, which took gloss into account, from their tristimulus values for the matte chips, they could actually create the hitching post themselves.

* Many art critics, when looking at Kelly's earlier *Spectrum* paintings, thought they were so precisely done, with each color being in exactly the right proportion to the other, that he must have used the Munsell system in painting them. He didn't.

† You've likely seen one of the CIE color spaces if you've ever poked around in your computer monitor settings and gone into the color-calibration menu. That weird rainbow guitar pick with a white starburst in the middle is one of the CIE color spaces, showing you the full gamut of colors reproducible by your monitor. You can really screw up your monitor by moving the center point of that white starburst. Try it! But don't call me to fix it.

But there was one more catch. The goal for any color standard was to use what's called color constant stimuli—that is, pigments or colorants that don't appear to radically change color with a change in lighting. This was difficult to do with gloss. That meant that for the Munsell glossy standard to actually work, the firm needed to very carefully choose and test and retest the pigments and suspensions and mixes of pigments for its chips to control the reflectance as much as they could so the standard would be standard.

The CIE XYZ math wasn't the hard part. It was this sourcing, mixing, measurement, calibration, remeasurement, and recalibration that would require a huge amount of work. Fortunately, they had a great lab assistant on hand to help.

There are three sets of data in the Davidson and Hemmendinger archive, working in concert with one another. There are pages for each Munsell hue (7.5YR, for instance), some typewritten, some handwritten, that are filled with different sets of measurements for each color chip that would show up on that hue page. One of the notations for each color chip they had to produce was an internal D&H product number; those product numbers correlate to sheaves of color samples on high-gloss paper stock, wide swipes of pigment that would make Ellsworth Kelly swoon with their spectral purity. Each one of these samples was run through a spectrophotometer to get the light-reflectance measurements; once the measurement was taken, it was compared with the Munsell notation. Additional sheets of formulas and measurements are stuffed into the back of the data set, noting which pigments were mixed and used in the creation of some of the samples, a quick tour through the history of paint manufacturers of the late 1950s—Sherwin-Williams Chrome Yellow, Tobey White, Duco #93. There are hundreds of pages of data, and nearly all the handwritten formulas, values, measurements, and notes are in Margaret's hand.[3]

Anyone who worked in color at this point in time would have told you that wasn't all that unusual. That's what lab assistants did, after all: the grunt work. But color had always made a place for women as equals. Davidson and Hemmendinger thought Margaret was such an important member of the team that she traveled with them to the 1957 winter meeting of the Optical Society of America—a crew that was used to seeing her with I.H.—to co-present a paper called "A Munsell Book in High-Gloss Colors."[4] Her name is nestled between Davidson's and Hemmendinger's: Rather than listing the lab assistant well after the names of main partners in the firm, they went in alphabetical order. It was Margaret's only technical paper, and her only scholarly citation.

The *Munsell Book of Color,* glossy edition, was released in 1958 and swiftly became the "master atlas" of Munsell colors, the standard for everything from plastics to car paint to glass. It still is, in fact: When you go to the Munsell website to buy a (spendy) copy of the *Munsell Book of Color,* the book you get offered is the Glossy Standard. When it was published, the Davidson and Hemmendinger firm was mentioned in the acknowledgments and thanked for their measurements and calculations. Margaret got an extra mention in the acknowledgments for actually producing the color chips for the book. But the chips are the most important part of the book. Without those chips—without Margaret's work—there'd be no Munsell glossy standard.

Davidson and Hemmendinger, thrifty from years of running their own business, naturally reused Margaret's work after she left the firm. Those notes on mixtures of color constant pigments, all that paint sourcing—could we automate that? Could we scale that up? And so it came to pass that in 1958, Davidson and Hemmendinger released COMIC, or Colorant Mixture Computer, an analog computer that could match and mix dyes or paints from a sample. You know about COMIC: Its descendants are the machines at the hardware store mixing up your gallon of Rejuvenate in satin finish.

Margaret had single-handedly defined the colors for one of the most talked-about dictionaries in American history and had been instrumental in the creation of one of the most used color standards around the world (and, by extension, the very first color-mixing computer in use); it's hard to believe that there could be any achievement that could top these. But there was one more small, personal jewel to be won.

In 1955, the original Problem 2 committee sans I.H. released the finished (mostly) ISCC-NBS standard. It contained Nickerson's paper from 1940 about how the standard was pulled together; a list of all the sources surveyed; charts for each hue divided up into "pockets" based on I.H.'s first "light/dark," "strong/weak" grid; and a comprehensive list of all color names used in their sources given by pocket. It did not take off: It was one of the only color standards in publication without a color chart. In 1958, Kenneth Kelly, encouraged by the release of the Munsell glossy standard, suggested that maybe it was time to pull together the color chips for the entire system—if not for the thousands of colors indexed in the standard, at least for the central colors for the 267 pockets. Only a few months later, in 1959, Problem 2 was resurrected by the ISCC with a view toward creating that visual standard, and Davidson and Hemmendinger were tapped to produce the chips.

Second verse, same as the first: Margaret, the practiced assistant, took the tristimulus and spectrophotometric measurements of each one of those 267 pocket colors and used the same system D&H had developed for the Munsell glossy standard in order to produce the glossy color chips for the ISCC-NBS. Sourcing the color-consistent paints was a cinch; she had done it just a year or two earlier for D&H. Like all things ISCC-NBS, funding came in staggered steps; the project was delayed and then picked up again and then delayed some more. While she was pulling the art books out for the grandkids to look at and perfecting her angel food cake, and throughout the great hullabaloo surrounding the release of the *Third,* she chipped away at the 267 color chips, which were finally produced in 1965 with the help of the Munsell Color Company and sold as a supplement to the ISCC-NBS standard. Each glossy chip was carefully placed on a hue page for each of the wedges of the Munsell color solid.

The final puzzle piece for the standard was neatly dropped into place thirty-four years later. In a pleasing piece of symmetry, Margaret was the one to complete the work that I.H. began right after that first ISCC meeting in 1931, years before the two of them even met.

Epilogue: Over the Rainbow

"One of the things I love about the ISCC-NBS," Maggie Maggio, a member of the ISCC Board of Directors, tells me, "is the massive amount of work that went into something that didn't go anywhere."[1]

I met Maggie at the annual ISCC conference in 2018. It was my first non-word conference, and I was an awkward jumble of tote bags* and new-kid jitters as I stood in line for registration. The ISCC is an interdisciplinary conference, so you never know whom you might meet. One person might make colorimeters. One person might teach colorimetry. One person might be on the color design team at General Motors. One may be an artist. One may be Leatrice Eiseman from Pantone. I was squirmy, because I was none of those things.

But nerds are nerds, no matter the field, and as soon as I told people why I was at the conference—Godlove, ISCC-NBS, "sea pink"—people were ready to talk color names, color perception, color language.† Once I told people what I did for a living—write

* This is my natural resting state, it should be noted.

† I knew I was at home when I walked into the ballroom for the keynote and found a group of folks examining the over-the-top rainbow decorations. They were mur-

dictionaries—they all leaned in and emphatically began talking color language.

If scholarship of the early twentieth century focused on the science of color creation and reproduction, the back half of the twentieth century swung inward and focused on how we see colors and whether names affect perception. In 1969, the anthropologist Brent Berlin and the linguist Paul Kay released a book-length analysis of our basic color names across a number of non-English languages and posited that, as languages (and the people who speak them) evolve, they add basic color names in a universal pattern, regardless of the language. The two basic color words that every language has are equivalent to the English "white" and "black." Then it moves up additively: The next color term is for red; the fourth color term added is for either yellow or green; the fifth term added is for whichever of the two fourth-stage color names they didn't add then; the sixth word added is for blue; the seventh is for brown; and if a language has eight or more basic color categories, then it adds the word for purple, pink, orange, or gray. They go even further and say that every language they studied also had about the same focal hue*—the center-of-gravity hue—for each color category: Using the Munsell system, they could prove, say, that the focal "green" is the same for everyone around the world.[2]

Berlin and Kay's theory of color-name evolution took the world by storm, and has been refined, argued with, built on, and torn down since. The idea that the names we give to colors affect our ability to perceive them was so entrancing that bits and pieces of

---

muring in a knot as I approached; I arrived just in time to hear one member of the group blurt out, "Khaki?? I didn't say 'khaki'! No, *tacky,* I said it was *tacky*! I know what 'khaki' is, sheesh."

* See the chapter "Down in Black and White" for a refresher course on focal hues, unique hues, and typical hues.

this research eventually made their way out into the broader world. The most common half-application is the case of the Himba people of Namibia; pop-sci articles will say that the Himba can't distinguish between blue and green because they have only four basic

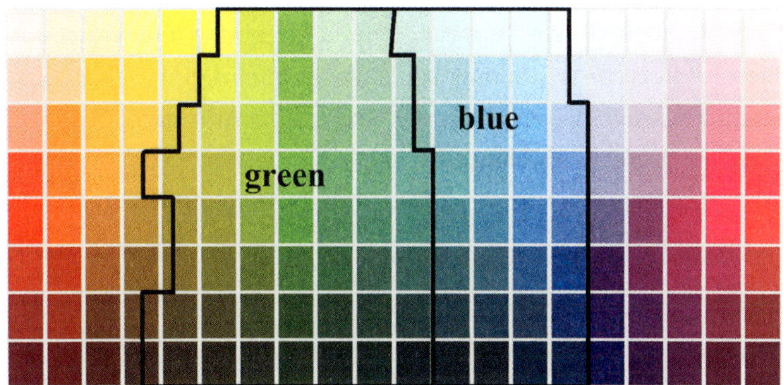

An array of Munsell hues with the colors that English speakers generally sort into the categories "green" and "blue" mapped onto it. You can thank Debi Roberson, Ian Davies, and Jules Davidoff (2000) for this map.

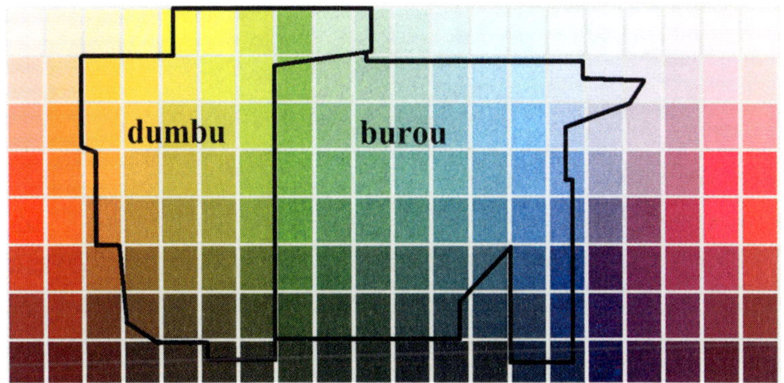

An array of Munsell hues with the colors Himba speakers generally sort into the categories *dumbu* and *burou* mapped out. You can also thank Debi Roberson, Ian Davies, and Jules Davidoff (2000) for this map.

color terms, and their last stage-four term is for green—no blue. This is a misreading and oversimplification of the study done on Himba color perception: The actual results are far cooler. First, the Himba actually have five main color terms, not four. And second, the Himba divide the blue-green part of the color space differently than English speakers do—and the color boundaries are pretty firm between the word they use for light greens and blues (*burou*) and the word they use for another group of colors that can include some light greens and blues (*dumbu*).

A BBC documentary crew that was doing a feature on the Himba got wind of this, and asked the anthropologists, Does this mean, then, that these colors are the same to the Himba?

Burou. Yes, both of them.

And the anthropologists confirmed that in Himba both of these colors are called *burou*.

This is the part of the study that we English speakers lose our minds over: *Aaaaah, they can't see blue!* But it's the wrong conclusion. It's not that the Himba can't see blue: Later studies show that they *can* separate out the bluer *burou* from the greener *burou*. It does take them longer, however, than identifying which of the squares on the facing page is *dumbu* and which is *burou*.

Aaaaah, English speakers can't see dumbu, the Himba now get to breathlessly claim.

Dumbu and *burou*. Obviously.

So yes, the Himba can see blue; the original study and later studies that confirm these findings seem to imply that there is a link between how we process color stimuli and how we name colors, and that is something far cooler than implying that this nomadic people aren't "evolved" enough to see blue.*

The real issue, of course, is whether we can somehow combine this whole language/perception thing with a standardized way of talking about or describing colors. While the ISCC-NBS was supposed to do that, other systems kept being created. As color monitors, web design, and the internet emerged, so, too, did the RGB color spaces and standards: HSL (for hue, saturation, and lightness), HSV (hue, saturation, and value), sRGB (standard Red, Green, Blue, and no, I don't know why "standard" isn't capitalized either), and the hexadecimal standard for web design (where #FFFFFF is white). Printers talk about CMYK (cyan, magenta, yellow, and key—

* You may be thinking to yourself, Haven't I seen charts from this study before? A circle of twelve green squares compared with a circle of eleven green squares and one sky-blue square? Why didn't you use *those* charts, Kory? Because those charts weren't actually used in the study: They were hastily put together for the benefit of the BBC documentary crew, who wanted some B-roll of the Himba taking a color discrimination test. Though, yes, those charts are very cool and get the point across well.

black). Scientists have expanded the CIE color space into CIELUV (for additive, RGB-type mixtures) and CIELAB (which translates between RGB and CMYK). Pantone is still around, too, and it now releases a Color of the Year. (The color for 2025: "Mocha Mousse," described on the Pantone site as "a warming, brown hue imbued with richness" that "nurtures us with its suggestion of the delectable qualities of chocolate and coffee, answering our desire for comfort."[3] Birren would approve, though it would cause Godlove to crumple his brow.) And Munsell, now owned by a company called X-Rite, is still used for food grading, in the evaluation of soils, and in the identification of rocks.

The multiplicity of systems was so much that even the stalwarts of the ISCC-NBS committee despaired. Kenneth Kelly and Deane Judd released a follow-up to the ISCC-NBS standard in 1976 that tried to accommodate all the many ways that we describe color. *Color: Universal Language and Dictionary of Names* suggests a six-level approach that lets the scientists keep their numbers but keys them to words for those of us who have an antagonistic relationship to math. But how helpful would that be in practice? The fact that neither it nor the ISCC-NBS itself has been widely adopted should tell you.

"You speak the language that you know," Maggio says, and despite the hint that maybe there could be some universal language, real-life applications favor a riot of proprietary systems.

So if it didn't work before, is there still time to make it work?

There is a huge debate brewing—simmering, possibly—about the language of color and color education. Take the primary colors. Why, *why,* in this day and age, when every specialist with even a smidge of color knowledge knows better, do we insist on giving kids red, blue, and yellow paint when everyone knows that the subtractive primaries—the three colors that, when mixed, give us black—are cyan, magenta, and yellow? Why can't Crayola *fix* this? Why can't *someone* fix this? A color researcher I had been chatting with turned to me. "You should just say, in your dictionaries, that the subtractive primaries are cyan, magenta, and yellow. You should just tell people."

Everyone else lit up—*another untried avenue for change!*—but I had to bear bad news. Linguists and lexicographers know, and will tell anyone who is listening, that no matter what you put in a dictionary, you cannot change common usage. "Cyan," "magenta," and "yellow," I explained, are all entered and defined in the *Third* as subtractive primaries, and that had clearly done diddly-squat vis-à-vis people's assumptions about primary colors. People had been trying for centuries to change common use in all sorts of ways by putting their thoughts, feelings, opinions, and even scientific facts in dictionaries. But established use is hard to contradict.

There are plenty of cases in the *Third* where the experts tried to change established use by pretending it didn't exist. "Pompadour" has a shocking amount of color-specific citational evidence in the Merriam-Webster files, considering that most of the colors entered in the *Third* have very, very little. There's a swatch from the Cheney Brothers' *Fabric Guide for Autumn 1926;* a reference slip from the *Oxford English Dictionary;* a 1915 citation from *The American Silk Journal;* a citation from the 1922 *Dictionary of English Phrases* about an English regiment called the Pompadours for the color of their uniforms; a citation from an antiquities catalog noting that Sèvres, a French porcelain manufacturer that was patronized regularly by Madame de Pompadour, created a pink ground for its porcelain in the eighteenth century and named it after their patroness. All these varied sources noted (or, in the case of the Cheney Brothers' citation, showed) "pompadour" was a pink. "Pompadour" was accordingly entered in *Webster's New International Dictionary* in 1909 as "a crimson or pink color."

But no, said I.H. (and the ISCC-NBS). "Pompadour" is a greenish blue, also known as pompadour green, and that was the only color that "pompadour" could be. Never mind that all the evidence in Merriam-Webster's files save one (a house-painting catalog) said that "pompadour" was some kind of pink, and that "pompadour" had been used to refer to some kind of pink across a number of industries since the mid-eighteenth century. The ISCC-NBS lists Maerz and Paul as the source; Maerz and Paul enter a "pompadour green" and cross-reference it to a color sample that, under daylight, certainly appears to be a medium blue and not a green or greenish blue at all. Hunting around in the Maerz and Paul index shows that they have marked this color as one "in constant daily and yearly use, that it is part of the ordinary vocabulary of color, and is not a seasonal or ephemeral term." This is true, though the "pompadour" that appears to actually be in constant daily and yearly use is not blue but pink.*

Did the ISCC-NBS's prescriptive declaration work? Do an image search online for "pompadour color swatch" and you'll see that the vast majority of commercial colors named "pompadour" are a rich reddish purple,† which is neither pink, blue, nor green. The definition in the *Third* stands, however, a testament to the struggles of standardization and the unyielding arrow of actual usage. I'm glad that's been sorted.

* So where on God's blue/green/pink earth did this whole notion of "pompadour" being a blue come from? Who knows. Maybe it had something to do with the Southeast Asian pompadour green pigeon, first illustrated by a Westerner in the mid-eighteenth century. The pigeon is a sage-green color; in the males, the mantle is a dark purple. But the pompadour green pigeon, as lovely as it is, is still not blue.

† Though one British paint manufacturer uses "pompadour" for a light baby blue. Always a contrarian in the bunch.

The *Third* stands—still, because the long-awaited new edition of the unabridged I began working on in 2010 is still not done, and likely never will be—but the color-defining style of the *Third* is long gone. The Merriam-Webster dictionaries abandoned the Godlovian defining formula almost as soon as they could. Starting with the seventh edition of the *Collegiate Dictionary* released in 1967, any color words that were slurped from the *Third* into the *Collegiate* lost the supplemental distinctions for space's sake. Many of the colors in the *Third* were never entered into the *Collegiate Dictionary:* Too many of them were now out of date or too rare to merit entry into an abridged, general dictionary. The modern definitions look a lot like Godlove's initial four-word definitions: a moderate red, a bright purplish pink, a deep reddish brown. Technology marches onward, too. Since the *Third* came out, we've invented the internet, and with just a quick search you can find examples of swatches for just about any color you can conceive of. The need for a print-only description of a color seems to be a distant memory.

We have also invented new colors and names for them—and keep inventing them. In just the last twenty years, we've gotten the super-black paint called Vantablack,* debuted in 2014, that seems to absorb all the light that hits it; in 2009, a scientist at Oregon State University accidentally synthesized a brand-new bright blue pigment that is more stable than other blue pigments, and the color was officially named YInMn Blue after the three chemicals used in its creation (though it's also informally known as Oregon Blue and Mas Blue after the scientist who created it);[4] in the late 1990s and

* Two fun tidbits about Vantablack. First, its name is based on the acronym for "vertically aligned nanotube arrays," and second, the paint manufacturer Surrey NanoSystems granted the artist Anish Kapoor exclusive rights to use Vantablack, meaning that any other super-black paints you may encounter that call themselves Vantablack are probably not Vantablack®.

early ’00s, the PhD student Andrew Parker, inspired by the types of iridescent yet velvety colors seen on butterflies and hummingbirds, mimicked the microscopic, transparent biological structures found on hummingbird feathers and butterfly wings and the ways that they refracted light, and created what he called "pure structural color";[5] in 2020, scientists at Purdue University created an ultra-white paint that reflects almost all the light that hits it, which means that it has a cooling effect on whatever it's applied to.[6] The ultra-white paint doesn't, as of this writing, have a name; nerds like me have been watching.

An unabridged dictionary is meant to cover as much of the language as possible, and to give as much definitional detail as possible for every word entered therein. But when it comes to describing colors, we seem to like a little vagueness. The relative paucity of complaints about the shortened color definitions in the *Merriam-Webster Collegiates* sure seems to hint that we have no problem accepting that "cerise," "strawberry," and "claret" are all "a moderate red." If we need to know how they're different, well, that's what paint chips (or color swatches or Pantone decks or field-specific color atlases) are for.

And this is, truly, the problem with any kind of standard, be it a color standard or a big lexical doorstop, and yet another tick on the "no" side of the column. The human mind is too expansive, too inventive, too frickin' squirrelly to be content with what is immediately in front of it. Colors in context speak to us in certain ways: When it comes to describing colors in a contextless void, it seems artificial—contrived, even. In that way, it's just like language: It becomes something worth watching only when it's about something outside itself. Our color names and their descriptions tend to be, as Birren would have said, felt. Which means "sea pink" is Cape Cod, a path lined with woodland green dotted through with sea pink, the irrational hot-cold feeling of blaze yellow sun and salt gray wind, the path of slate and buff and umber, a plain of vista blue melting into azure, an Ellsworth line of hue where they meet in the middle—interrupted by the blaze orange cries of children by a maroon square edged in lobster, a figure clothed in seashell, a proffered Swiss coffee swirl on an amber cone, a beacon punctuating a delightfully flambic afternoon.

Tell me you don't see that exactly as I do.

Acknowledgments

This book was about ten years in the making from soup to nuts—longer if you go back to the first blog post I wrote about the color definitions in the *Third*. Consequently, there are many, many people to thank for their time and expertise. Buckle up.

This book wouldn't be possible—or, frankly, coherent—without the many members of the ISCC who were willing to let me hang out with them and ask questions, or who taught me about color science and color measurement (sometimes without them even knowing they were doing it). Many thanks to Maggie Maggio, Dave Wyble, John Seymour, Paul Green-Armytage, Luanne Stovall, Dimitris Mylonas, Amy Woolf, David Briggs, Mark Fairchild, Peter Donahue, Harald Arnkil, and Berit Bergström.

Though this book is about color, it's also about dictionaries, so heartfelt thanks to Steve Kleinedler for his *American Heritage* knowledge and linguistics suggestions, Krista Williams for her research on dictionaries and color terms, and Emily Brewster for her smarts and the extra bed whenever I visited the office.

Andrew Miller, my first editor at Knopf, and Maris Dyer, my second editor at Knopf, are owed as much beer as they could ever desire for enduring all the drafts, rewrites, delays. Andrew, especially: He recognized that a chapter I submitted for *Word by Word* was actually the start of this book, and encouraged me to write it.

Heather Schroder, my agent at Compass Literary, ditto on all counts. Always in my corner, she continues to encourage me to think like a writer, be a writer, do what it takes to write, no matter what the circumstances.

I may have written this book, but it wouldn't exist without the hard work of everyone at Knopf who took my blah bitty blah and turned it into this marvelous book/audiobook/e-book you currently hold in your hot little hands: Amy Brosey, Chip Kidd, Lisa Montebello, Emily Murphy, Cassandra Pappas, Elka Roderick, and Ingrid Sterner.

Ansa Stamper provided the raccoon illustration that makes the Ogden-Richards triangle of meaning understandable *and* adorable.

No writer or researcher ever "discovers" something in a professionally held or managed archive, just like Christopher Columbus didn't "discover" America and lexicographers don't "discover" new words. I owe 95 percent of this book to the archivists, accessions specialists, reference librarians, and library assistants who gathered, organized, conserved, created finding aids for, occasionally digitized, and housed the papers I sifted through between 2017 and 2020:

At the Hagley Library & Museum, I would like to thank Lucas Clawson, former Reference Archivist and Hagley Historian; Cheryl Jackson, former Library Coordinator; Clayton Ruminski, former Archives Specialist; and John Waters, former Stack Assistant, for their help in accessing so many of their collections (and not panicking or kicking me out when I hollered in the archives room that one time).

At the American Heritage Center at the University of Wyoming, I'd like to thank Molly Marcusse, former Reference Archivist; John Waggener, former Reference Archivist and current University Archivist and Historian; Ginny Kilander, Reference Services Manager; and Vicki Glanz, former Archives Assistant and current Reference Services Coordinator, for their patience as I camped out in the archives reading room with Gove's papers, diligently taking notes and cellphone photos (and swearing quietly when those photos were just of my lap).

At Smith College's Special Collections, where I read William Allan Neilson's papers while the library that bore his name was under renovation, I'd like to thank Amy Hague, the now-retired Curator of Manuscripts and Research Services Archivist, and Mary Biddle, former Reading Room Supervisor, for giving me a warm Smithie welcome despite the fact that they had to haul Neilson's papers out of storage and hump them up to the second floor of Bass Hall so I could flip through them. While none of his collected papers are quoted in the published version of this book, his papers were instrumental in setting the scene for Gove's entry.

At the Houghton Library at Harvard University, where I accessed James Parton's papers and found some good information about the creation of the *American Heritage Dictionary* (though none of it is quoted in the published version of this book), my very best regards to Susan Halpern, who not only guided me through the arcane rituals of getting a reading card for the Houghton Library, but pulled Parton's papers for me and set me up at a very cushy table in the reading room for several days.

And though they are not formal collections in the sense that there's no finding aid, no accessioning archivist, and, in one case, only one dim bulb illuminating the entire collection, I would be remiss in not thanking the holders of the two private collections I was also given access to.

For the collection of letters that started this whole mess: Many thanks to the generations of secretaries, typists, and editorial librarians (most of them women, of course) at Merriam-Webster who dutifully filed all the Godlove, Carhart, Oakes, and Gove correspondence in orderly array so I could one day stumble across *one letter* and be able to find the rest. Also to Matthew Dube at Merriam-Webster for helping me fill in research gaps after I left the company in 2018 (and for agreeing that the single-bulb fluorescent in the hallway where the W3 correspondence was housed was, truly, not ideal) and to E. Ward Gilman, Director of Defining emeritus, who not only answered every question I peppered him with when I was a baby lexicographer at Merriam-Webster, but did the same after I left and peppered him with questions about Gove, Driscoll, W3, proofreading, ad infinitum. He passed away in 2022. I miss him.

And for the collection of letters that provided the glue for the book and really introduced me to Margaret: Deepest heartfelt thanks to Terry Godlove Jr. and Karen Malley for sharing memories and laughs over lunch, and for giving me access to their grandparents' papers. It's no small thing to let some random writer sit in your dining room and leaf through the record of two lives; no small thing to trust your grandparents' story to a relative stranger. May your kindness come back to you tenfold. Thanks for sharing your grandparents with me.

Deepest gratitude to and for Josh, my own personal prism.

Notes

All dictionary definitions are taken from *Webster's Third New International Dictionary of the English Language, Unabridged,* unless otherwise stated.

Archives and Collections

DNP: Dorothy Nickerson Papers (2188-V), Manuscripts and Archives, Hagley Museum & Library, Wilmington, Del., 19807

GFA: Godlove (Margaret and I. H.) Papers, Godlove Family Archive, privately held

HHP: Henry Hemmendinger Papers (2325), Manuscripts and Archives, Hagley Museum & Library, Wilmington, Del., 19807

ISCC: Inter-Society Color Council Records (2188), Manuscripts and Archives, Hagley Museum & Library, Wilmington, Del., 19807

MWCA: Merriam-Webster corporate in-house archives, Merriam-Webster, Springfield, Mass., 01105

PBGP: Philip Babcock Gove Papers (5901), American Heritage Center, University of Wyoming, Laramie, Wyo., 82071

DYEING KILLS

1. Science History Institute, "Justus von Liebig and Friedrich Wöhler."
2. *Britannica,* "synthetic dyes," www.britannica.com.
3. Canadian War Museum, "Weapons on Land—Poison Gas."
4. American Chemical Society, "2,4,6-Trinitrotoluene."
5. Haskin, "Dye Problem."
6. Clark, *Dyeing for a Living,* 12.
7. G. T. M., "Dye Famine in America and the Proposed Remedy." The same author assured the reader that "the existing shortage would soon disappear." The article this appears in was published in December 1915; the Dye Famine would last well into the 1920s.
8. Cochrane, *Measures for Progress,* 190.

9. National Institute of Science and Technology, "Founding: Overview."
10. Office of Weights and Measures, "OWM Background and History."
11. Quoted in Cochrane, *Measures for Progress,* 171.

THE YELLOW BRICK ROAD

1. U.S. Committee on Agriculture and Forestry, *Testimony in Regard to the Manufacture and Sale of Imitation Dairy Products,* 95.
2. Ibid., 76–77.
3. Peek, "I Can't Believe It's Not Yellow."
4. Syme, *Werner's Nomenclature of Colours,* 37.
5. Munsell, "Munsell Diary," Nov. 27, 1900.
6. Ibid.
7. Rosa, "Expenditures and Revenues of the Federal Government," 105–8.

OUT OF THE BLUE

1. "Funeral of Homer Merriam," *Transcript-Telegram* (Holyoke, Mass.), June 25, 1908, 9.
2. Carhart to Priest, Nov. 14, 1929, and Priest to Carhart, Nov. 30, 1929, MWCA.
3. Priest to Carhart, Nov. 30, 1929, MWCA.
4. Ibid.
5. Johnson, "Pink, n.s.," in *Dictionary of the English Language.*
6. "Puce," in Whitney, *Century Dictionary,* pt. 16.
7. Quotation and definitions from Priest to Carhart, July 30, 1930, MWCA.
8. Maerz and Paul, preface to *Dictionary of Color,* v.
9. Quotation and definitions from Priest to Carhart, July 30, 1930, MWCA.
10. Ibid.
11. Carhart to Priest, Aug. 1, 1930, MWCA.
12. Priest to Carhart, Oct. 22, 1930, MWCA.

THE GOLDEN BOY

1. Excerpted from "He Couldn't Escape Color," *Rainbow: The Employee Magazine of General Aniline and Film,* 1945; ISCC *Newsletter,* no. 62 (Nov. 1945), 2188-I.-E., Newsletters, ISCC.
2. Undated, GFA.
3. B., "Exhibition on the Science and Art of Color."
4. I. H. Godlove, Curriculum vitae, n.d., GFA.
5. Both the type-hollering and the resignation are taken from Priest to Carhart, Nov. 27, 1930, MWCA.
6. Carhart to Godlove, Jan. 17, 1931, MWCA.
7. Coinage and footnoted explanation that, whatever, it's *fine,* people will get over it, in Priest to Carhart, Feb. 2, 1931, MWCA.
8. Godlove to Carhart, March 20, 1931, MWCA.
9. Godlove to Carhart, March 28, 1931, MWCA.
10. Ibid.

A BROWN STUDY

1. Carhart to Godlove, May 26, 1931, MWCA.
2. Godlove to Carhart, June 7, 1931, MWCA.
3. Quoted in Skinner, *Story of Ain't.*
4. Baker to Carhart, interoffice memo, June 11, 1931, MWCA.
5. Ibid.
6. Annotation, ibid.
7. Godlove to Carhart, Nov. 17, 1931, MWCA.
8. Godlove to Bittinger (copy), March 20, 1932, MWCA.
9. Godlove to Carhart, July 16, 1932, MWCA.
10. "Phonetics Expert Takes Own Life," *Brattleboro (Vt.) Reformer,* Oct. 27, 1933, 8.
11. Godlove to Merriam-Webster, Nov. 7, 1933, MWCA.

THE SILVER BULLET

1. Burton, "Color Wizards Identify 100,000 Hues by Eye."
2. Crider, "Unpuzzling Color."
3. These examples were pulled from Godlove, "Progress Report of the Committee on Measurement and Specification," Inter-Society Color Council, *Bulletin,* June 7, 1932, 2188-I.-B.-1., ISCC.
4. "Color Nomenclature in the United States Pharmacopoeia," ISCC *Newsletter,* Oct. 16, 1933, 2188-I.-E., ISCC.
5. Priest to Carhart, July 30, 1930, MWCA.
6. "Progress Report of the Committee on Measurement and Specification," Inter-Society Color Council, *Bulletin,* June 7, 1932, 2188-I.-B.-1., ISCC.
7. Baptista and Travis, "I. G. Farben in America."
8. "Synchronized Men," ISCC *Newsletter,* no. 60 (July 1945); "Delegate Scofield," ISCC *Newsletter,* no. 59 (May 1945); "Orange a Shade of Yellow?," ISCC *Newsletter,* no. 58 (March 1945); "Colors for Street Cars," ISCC *Newsletter,* no. 62 (Nov. 1945). All found in 2188-I.-E., ISCC.
9. Godlove to Knott, April 14, 1945, MWCA.

RED TAPE

1. Godlove to Knott, April 14, 1945, MWCA.
2. Memorandum to the Editorial Department, April 8, 1949, MWCA.
3. "Justice Cardoza Among Those Williams Honors," *North Adams (Mass.) Evening Transcript,* June 20, 1932, 8.
4. Whitney to John Johnson, July 29, 1918, quoted in Reingold and Reingold, *Science in America.*
5. "The History of Popular Science," *Popular Science,* July 24, 2002, www.popsci.com.
6. Oakes to Editorial Board, in-house memo, July 24, 1944, MWCA.
7. Oakes to Bethel, in-house memo, April 25, 1945, MWCA.
8. Ibid.
9. Oakes, draft of M-W answer to Godlove, April 25, 1945, MWCA.
10. Ibid.

THE GRAY HEAD

1. "Minutes of Consultation with Dr. P. W. Long," Oct. 25, 1946, 5901:1, PBGP.
2. Munroe to Long, Oct. 16, 1946, 5901:6, PBGP.
3. "List of Displaced German Scholars and Supplementary List of Displaced German Scholars" (1936), Bulmash Family Holocaust Collection, 2022.1.12ab, digital.kenyon .edu.
4. Oakes to Bethel, confidential in-house memo, Nov. 9, 1946, 5901:6, PBGP.
5. Oakes to Bethel, confidential in-house memo, Dec. 13, 1946, 5901:6, PBGP.
6. Oakes and Bethel to Munroe, confidential in-house memo, Nov. 27, 1946, 5901:6, PBGP.
7. Footnote quoted in Bender to Munroe, confidential letter, Jan. 29, 1947, 5901:6, PBGP.
8. Oakes to Bethel, confidential in-house memo, Aug. 22, 1947, MWCA.
9. Bethel and Holt to Munroe, in-house memo, Aug. 29, 1946, MWCA.
10. Oakes to Munroe, Aug. 30, 1946, MWCA.
11. Twaddell to Bethel, confidential in-house memo, April 20, 1951, 5901:6, PBGP.
12. Munroe to Gove, offer letter, June 27, 1951, 5901:6, PBGP.

THE GREENHORN

1. Gove to Merriam-Webster, Feb. 12, 1946, 5901:6, PBGP.
2. Quoted in Morton, *Story of Webster's Third,* 33.
3. Oakes, "Air force" [draft], Jan. 7, 1949, MWCA.
4. Gove, memo at "air force," Jan. 13, 1949, MWCA.
5. "Minutes of a Meeting of the Editorial Board of G & C Merriam Co.," Nov. 21, 1951, MWCA.
6. Ibid.
7. Ibid.
8. Ibid.
9. "Minutes of a Meeting of the Editorial Board of G & C Merriam Co.," Feb. 18, 1953, MWCA.
10. E. Ward Gilman, interview by Emily Brewster, Sept. 22, 2012, MWCA.
11. Oakes to Gove and Bethel, memo regarding etymology for the *Third,* Nov. 29, 1951, MWCA.
12. "Reports of Office Queries," Sept. 22, 1952, MWCA.
13. Oakes to Bethel and Gove, and Gove back to Oakes, in-house memo, Dec. 12, 1951, MWCA.

DOWN IN BLACK AND WHITE

1. Godlove to Bethel, Sept. 21, 1948, MWCA.
2. Godlove to Oakes, June 2, 1950, MWCA.
3. Oakes to Godlove, June 9, 1950, MWCA.
4. "Typical ISCC-NBS Color Descriptions," attachment, Godlove to Oakes, June 2, 1950, MWCA.
5. Ibid.
6. "Special Editors," in *Webster's New International Dictionary, Second Edition,* xviii.
7. Oakes to Bethel, in-house memo, March 9, 1949, MWCA.

IN LIVING COLOR

1. "Puzzlers Open 103rd Session Here by Recognizing 45-Letter Word," *New York Herald Tribune,* Feb. 23, 1935.
2. Oakes to Godlove, Nov. 20, 1952, MWCA.
3. Skorinko et al., "Rose by Any Other Name."
4. "TCCA Activities," ISCC *Newsletter,* no. 99 (March 1952), 2188-I.-E., ISCC.
5. Blaszczyk, *Color Revolution,* 155.
6. Lois Long, "On and Off the Avenue," *New Yorker,* Nov. 3, 1956, 176.
7. "Helena Rubinstein Announces Sensational Silken Lipstick!," advertisement.
8. Birren, *Selling with Color,* 40.
9. Eiseman, *Color,* 69.
10. Herbert and Mead, *King of Color,* 184–85.
11. Color cards for Mercury's Ditzler paints for 1951, 1952, and 1953, courtesy of paintref.com, which catalogs paints by color name, manufacturer, manufacturer code, and year for classic cars. Hot-rodders, get thee hence.
12. "Fall Bonnets," *Harper's Bazaar,* Sept. 23, 1876, 611.
13. "Sale of 'Nigger' Brown Dress Goods," ad, *Tennessean* (Nashville), Nov. 29, 1914, 7.
14. "Nigger Brown," *New York Age,* Dec. 11, 1913, 4.
15. "The Fashionable Color for the Coming Season," *World* (Coos Bay, Ore.), Sept. 12, 1914, 11.
16. Godlove to Granville, copy attached to Godlove to Oakes, Jan. 10, 1953, MWCA.
17. Oakes to Godlove, Nov. 20, 1952, MWCA.
18. Godlove to Oakes, Nov. 23, 1952, MWCA.
19. Ibid.
20. Ibid.

RED HOT

1. Memo from Asa Baker to Thomas Knott, March 10, 1934, MWCA. "If we could put these figures into the Introduction somewhere, would it not be good dope?" writes Baker, a man who probably never actually had uttered the phrase "good dope" in his life.
2. "Report on Third Edition Progress," Nov. 10, 1952, 5901:6, PBGP.
3. Oakes to Godlove, May 22, 1953, MWCA.
4. Godlove to Oakes, May 23, 1953, MWCA.
5. Godlove to Oakes, Oct. 2, 1953, MWCA.
6. "Motor Neuron Diseases," National Institute of Neurological Disorders and Stroke, April 14, 2024, www.ninds.nih.gov.
7. Edward Oakes, "Color," in-house style guide for the *Third,* May 24, 1955, MWCA.
8. Gove to Gallan, Feb. 24, 1953, 5901:6, PBGP.
9. Godlove to Oakes, Oct. 8, 1953, MWCA.
10. Kelly to Godlove, copy enclosed in Godlove to Oakes, July 8, 1953, MWCA.
11. Oakes to Gove, in-house note, June 16, 1954, MWCA.
12. Margaret Godlove to Oakes, Aug. 14, 1954, MWCA.
13. Oakes to Margaret Godlove, Aug. 16, 1954, GFA.

THE GOLDEN GIRL

1. Rossiter, *Women Scientists in America: Struggles and Strategies to 1940.*
2. "Oberlin College—125th Anniversary Alumni Catalog," Margaret Noss Godlove, Autumn 1958, RG 20, Alumni Association Records, 1839–present, Oberlin College Archives, Oberlin College.
3. "Calling Boston, Chicago, and Hollywood," ISCC *Newsletter,* no. 62 (Nov. 1945), 2188-I.E., ISCC.
4. Smith, "Health Hazards of the Dye Industry."
5. "Reminiscences of General Aniline and Film Corporation by Harlan B. Freyermuth."
6. "Recent Activities of the TCCA," ISCC *Newsletter,* no. 55 (Sept. 1944), 2188-I.E., ISCC.
7. Florence Wall to Margaret Godlove, n.d., GFA.
8. Miriam Hemmendinger to Margaret Godlove, Aug. 17, 1954, GFA.
9. Glenda Rose to Margaret Godlove, Aug. 24, 1954, GFA.
10. Kelly to Margaret Godlove, Aug. 23, 1954, GFA.
11. "Necrology: I. H. Godlove."
12. Harold Steigler to Margaret Godlove, Aug. 23, 1954, GFA.
13. Nickerson to Margaret Godlove, Aug. 17, 1954, GFA.
14. "Foreword," ISCC *Newsletter,* no. 115 (Jubilee Issue) (Nov. 1954), 2188-I.E., ISCC.
15. Esther [no last name given] to Margaret Godlove, n.d., GFA.

PINK COLLAR

1. "Chrome green," draft definition, Feb. 1954, MWCA.
2. Oakes note to entry for "chrome green," June 15, 1955, MWCA.
3. Godlove to Oakes, May 29, July 26 and 31, 1953, MWCA.
4. Oakes to Gove, in-house memo, May 19, 1954, MWCA.
5. Oakes to M. Godlove, Sept. 15, 1954, MWCA.
6. "The Big Book," n.d., 5901:16, PBGP.
7. Russell, *Women and Dictionary-Making,* 172.

THE END OF THE RAINBOW

1. "Blue," second draft for the *Third,* March 26, 1955, MWCA.
2. "Report of Third Edition Progress (7th report)," June 15, 1955, 5901:6, PBGP.
3. Oakes to Gove, March 5, 1956, MWCA. Oakes's resignation is on stationery from home and was sent from home, though he was very much still in the office. Already he had begun to disentangle himself from Merriam.
4. Gove to Oakes, March 8, 1956, MWCA. Gove's response is on company letterhead, sent to Oakes's home address. No collegial desk visits or closed-door meetings in the office for Gove.
5. Oakes to Gove, March 11, 1956, MWCA.
6. Gallan to Oakes, June 20, 1957, MWCA.
7. "Color," draft definition for the *Third,* March 28, 1957, MWCA.
8. "Color, n," *Webster's Third New International Dictionary, Unabridged,* 1961.

THE COLOR OF MONEY

1. "Report of Third Edition Progress (15th Report)," June 10, 1959, 5901:6, PBGP.
2. "Webster's New Word Book," editorial, *New York Times*, Oct. 12, 1961, 21.
3. "Expert Says 'Necking' Good Word," *Raleigh (N.C.) News & Observer*, July 11, 1964, clipping in 5901:15, PBGP.
4. Corrigan, "Dictionary Defended by Merriam Editor," 20.
5. Alice Hughes, "Verve and Zing Added to the New Dictionary," *Poughkeepsie (N.Y.) Journal*, Sept. 13, 1961, 6.
6. Graham Dushane, "'Say It *Ain't* So!,'" *Science*, Nov. 10, 1961, 1493, doi.org/10.1126/science.134.3489.1493.
7. "Winners and Sinners," *New York Times*, Jan. 4, 1962, 122–23.
8. Follett, "Sabotage in Springfield."
9. Macdonald, "String Untuned." Please note the extra-sly use of "revolution" in his review—an echo of all the "revolutionarys" of the press release.
10. Barzun, "What Is a Dictionary?"
11. Quoted in Jewett, *Science Under Fire*, 80.
12. Ibid.
13. Barzun, *Science*, 201.
14. Laurance Hart, "Review of Webster's Third New International Dictionary," *Lexiconoclast*, March 15, 1962, MWCA. Hart was a civil engineer; you'd think that he'd have enjoyed all that highly technical encyclopedic stuff.
15. Nickerson to Gallan, March 26, 1962, MWCA.
16. "Firmament blue," draft defining slip for the *American Heritage Dictionary*, n.d., 2188-V.-B.-11, DNP.
17. "Eggplant," draft defining slip for the *American Heritage Dictionary*, n.d., 2188-V.-B.-11, DNP.
18. "Eggplant," draft defining slip for the *Third*, Feb. 4, 1954, MWCA.

THE GOLD STANDARD

1. Quoted in Nickerson, "History of the Munsell Color Foundation," 9.
2. "Oberlin College—125th Anniversary Alumni Catalog."
3. Color Samples, n.d., 2325-V.-A., HHP. While very few of these are dated, the annotations and accompanying notes match Margaret's handwriting (which I compared with what's in MWCA, DNP, and GFA).
4. "Program of the 1957 Winter Meeting of the Optical Society of America."

EPILOGUE: OVER THE RAINBOW

1. All quotations from Maggio are from Maggie Maggio, interview by author, June 29, 2019.
2. Berlin and Kay, *Basic Color Terms*.
3. Pantone, "PANTONE 17-1230 Mocha Mousse | Pantone Color of the Year 2025."
4. "YInMn Blue," Oregon State University, Department of Chemistry, Oct. 5, 2022, chemistry.oregonstate.edu.
5. "Crimson Pure Structural Colour® Disc | Sage Culture," Sage Culture, accessed June 1, 2024, sageculture.com.
6. Purdue University News Service, "The Whitest Paint Is Here—and It's the Coolest. Literally," April 15, 2021, www.purdue.edu.

Bibliography

Adelson, Rachel. "Hues and Views." *Monitor on Psychology* 36, no. 2 (2005). www.apa.org.

Albers, Josef. *Interaction of Color*. 4th ed. Yale University Press, 2013.

American Chemical Society. "2,4,6-Trinitrotoluene." Accessed May 30, 2024. www.acs.org.

American Society for Testing Materials and Inter-Society Color Council. *Symposium on Color: Its Specification and Use in Evaluating the Appearance of Materials*. ASTM International, 1941.

B. "Exhibition on the Science and Art of Color." *Science* 73, no. 1881 (1931): 69–70.

Babin, Barry J., David M. Hardesty, and Tracy A. Suter. "Color and Shopping Intentions." *Journal of Business Research* 56, no. 7 (2003): 541–51. doi.org/10.1016/s0148-2963(01)00246-6.

Ball, Philip, and Mario Ruben. "Color Theory in Science and Art: Ostwald and the Bauhaus." *Angewandte Chemie International Edition* 43, no. 37 (2004): 4842–47. doi.org/10.1002/anie.200430086.

Baptista, Robert, and Anthony Travis. "I. G. Farben in America: The Technologies of General Aniline & Film." *History and Technology* 22, no. 2 (2006): 187–224.

Barzun, Jacques. *Science: The Glorious Entertainment*. Harper & Row, 1964.

———. "What Is a Dictionary?" *American Scholar* 32, no. 2 (1963): 176–81.

Berlin, Brent, and Paul Kay. *Basic Color Terms: Their Universality and Evolution*. University of California Press, 1969.

Biggam, C. P. *The Semantics of Colour: A Historical Approach*. Cambridge University Press, 2015.

Bird, C. M., S. C. Berens, A. J. Horner, and A. Franklin. "Categorical Encoding of Color in the Brain." *Proceedings of the National Academy of Sciences* 111, no. 12 (2014): 4590–95. doi.org/10.1073/pnas.1315275111.

Birren, Faber. *Color Perception in Art*. Van Nostrand Reinhold, 1976.

———. *Color Psychology and Color Therapy: A Factual Study of the Influence of Color on Human Life*. Citadel Press, 1950.

———. *Selling with Color*. McGraw-Hill, 1945.

Blaszczyk, Regina Lee. *The Color Revolution*. MIT Press in Association with the Lemelson Center, Smithsonian Institution, 2012.

Britton, W. Earl. "Some Effects of Science and Technology upon Our Language." *College Composition and Communication* 21, no. 5 (1970): 342–46. doi.org/10.2307/356083.

Burton, Walter E. "Color Wizards Identify 100,000 Hues by Eye." *Popular Science*, Dec. 1935.

Canadian War Museum. "Weapons on Land—Poison Gas." Canada and the First World War. Accessed May 30, 2024. www.warmuseum.ca.

Chevreul, Michel-Eugène. *The Principles of Harmony and Contrast of Colors, and Their Applications to the Arts*. Rev. ed. Schiffer, 1987.

Clark, Mark. *Dyeing for a Living: A History of the American Association of Textile Chemists and Colorists, 1921–1996*. American Association of Textile Chemists and Colorists, 2001.

Cochrane, Rexmond C. *Measures for Progress: A History of the National Bureau of Standards*. U.S. Department of Commerce, National Bureau of Standards, 1974.

Corrigan, Faith. "Dictionary Defended by Merriam Editor." *Cleveland Plain Dealer*, April 21, 1962.

Crider, John H. "Unpuzzling Color." *Scientific American* 158, no. 5 (1938): 284–85. doi.org/10.1038/scientificamerican0538-284.

Davidson, H. R., and Henry Hemmendinger. "Colorimetric Calibration of Colorant Systems." *Journal of the Optical Society of America* 45, no. 3 (1955): 216–19. doi.org/10.1364/josa.45.000216.

Douglas, Lamb. "A Brief History of Automotive Coatings Technology." American Coatings Association. Accessed Jan. 1, 2024. www.paint.org.

Dutton, William S. *DuPont: One Hundred and Forty Years*. Charles Scribner's Sons, 1951.

"Dye—Synthetic, Organic, Colorants." In *Encyclopaedia Britannica*. Accessed May 30, 2024. www.britannica.com.

Eiseman, Leatrice. *Color: Messages & Meanings*. Hand, 2007.

———. *Pantone: The 20th Century in Color*. Chronicle, 2011.

FabricMate. "Whisper—Panel and Acoustic Fabric." Sales brochure, 2019.

Follett, Wilson. "Sabotage in Springfield: Webster's Third Edition." *Atlantic Monthly*, Jan. 1962.

Franklin, A., G. V. Drivonikou, A. Clifford, P. Kay, T. Regier, and I. R. L. Davies. "Lateralization of Categorical Perception of Color Changes with Color Term Acquisition." *Proceedings of the National Academy of Sciences* 105, no. 47 (2008): 18221–25. doi.org/10.1073/pnas.0809952105.

Frawley, William. "Lexicography and the Philosophy of Science." *Dictionaries: Journal of the Dictionary Society of North America* 3, no. 1 (1980): 18–27. doi.org/10.1353/dic.1980.0016.

G. & C. Merriam Co. *The House That Merriam-Webster Built*. G. & C. Merriam, 1930.

Gibson, Edward, Richard Futrell, Julian Jara-Ettinger, Kyle Mahowald, Leon Bergen, Sivalogeswaran Ratnasingam, Mitchell Gibson, Steven T. Piantadosi, and Bevil R. Conway. "Color Naming Across Languages Reflects Color Use." *Proceedings of the National Academy of Sciences* 114, no. 40 (2017): 10785–90. doi.org/10.1073/pnas.1619666114.

Godlove, Isaac Hahn. *The Earliest Peoples and Their Colors*. Unpublished manuscript fragment, n.d.

Gove, Philip Babcock. "Letter to the Editor." *New York Times*, Nov. 5, 1961.

———, ed. *Webster's Third New International Dictionary of the English Language, Unabridged*. 1961. Merriam-Webster, 2002.

Green-Armytage, Paul. "Colour, Language, and Design." PhD diss., University of Western Australia, 2005.

G. T. M. "The Dye Famine in America and the Proposed Remedy." *Nature* 96, no. 2407 (1915): 429–30. doi.org/10.1038/096429a0.

Haskin, Frederic J. "The Dye Problem." *News Journal*, Feb. 10, 1916.

Hemmendinger, David. "COMIC: An Analog Computer in the Colorant Industry." *IEEE Annals of the History of Computing* 36, no. 3 (2014): 4–18. doi.org/10.1109/mahc.2014.34.

Herbert, Lawrence, and Linda Mead. *The King of Color*. Pantone, 2007.

Howard, Ian P. "Guest Editorial: The Helmholtz-Hering Debate in Retrospect." *Perception* 28 (1999): 543–49.

Hull, Callie, ed. *Industrial Research Laboratories of the United States, Including Consulting Research Laboratories*. 8th ed. Bulletin of the National Research Council 113. National Academy of Sciences, 1946. doi.org/10.17226/18663.

Inter-Society Color Council. *Comparative List of Color Terms*. Inter-Society Color Council, 1949.

"ISCC-NBS Centroid Color Charts." Inter-Society Color Council and National Bureau of Standards, 1958.

Itten, Johannes. *The Art of Color: The Subjective Experience and Objective Rationale of Color*. Reinhold, 1961.

Jewett, Andrew. *Science Under Fire: Challenges to Scientific Authority in Modern America*. Harvard University Press, 2020.

Johnson, Samuel. *A Dictionary of the English Language*. 1755. johnsonsdictionaryonline.com.

Judd, Deane B. *Color in Our Daily Lives*. U.S. Department of Commerce, National Bureau of Standards, 1975.

Judd, Deane B., and Kenneth L. Kelly. "Method of Designating Colors." *Journal of Research of the National Bureau of Standards* 23, no. 3 (1939): 355–85. doi.org/10.6028/jres.023.019.

Kay, P., and T. Regier. "Resolving the Question of Color Naming Universals." *Proceedings of the National Academy of Sciences* 100, no. 15 (2003): 9085–89. doi.org/10.1073/pnas.1532837100.

Kelly, Kenneth L. *Coordinated Color Identifications for Industry*. NBS Technical Note 152. U.S. Department of Commerce, National Bureau of Standards, 1962.

Kelly, Kenneth L., and Deane B. Judd. *Color: Universal Language and Dictionary of Names*. NBS Special Publication 440. U.S. Department of Commerce, National Bureau of Standards, 1976.

———. *The ISCC-NBS Method of Designating Colors and a Dictionary of Color Names*. National Bureau of Standards Circular 553. U.S. Department of Commerce, 1955.

Landau, Sidney I. *Dictionaries: The Art and Craft of Lexicography*. 2nd ed. Cambridge University Press, 2014.

Liberman, Mark. "Himba Color Perception." *Language Log* (blog), March 17, 2015. languagelog.ldc.upenn.edu.

Livingstone, Margaret. *Vision and Art: The Biology of Seeing*. Abrams, 2014.

Lynch, Jack. *You Could Look It Up*. Bloomsbury Publishing USA, 2016.

Macdonald, Dwight. "The String Untuned." *New Yorker,* March 10, 1962.

Maerz, Aloys John, and M. Rea Paul. *A Dictionary of Color*. McGraw-Hill, 1930.

McDowell, Clyde S. "Naval Research." *United States Naval Institute Proceedings* 45 (1919).

Miller, Elizabeth G., and Barbara E. Kahn. "Shades of Meaning: The Effect of Color and Flavor Names on Consumer Choice." *Journal of Consumer Research* 32, no. 1 (2005): 86–92. doi.org/10.1086/429602.

Moroney, Nathan. "Unconstrained Web-Based Color Naming Experiment." In *Color Imaging VIII: Processing, Hardcopy, and Applications.* Edited by Reiner Eschbach and Gabriel G. Marcu. SPIE, 2003. doi.org/10.1117/12.472013.

Morton, Herbert Charles. *The Story of Webster's Third: Philip Gove's Controversial Dictionary and Its Critics*. Cambridge University Press, 1995.

Munsell, Albert. *A Color Notation*. Geo. H. Ellis, 1905.

———. "Munsell Diary." Educational Resources, Munsell Color Science Laboratory. www.rit.edu.

Munsell Color Company. "The Difference Between Chroma and Saturation," June 24, 2016. munsell.com.

———. *Munsell Book of Color: Defining, Explaining, and Illustrating the Fundamental Characteristics of Color*. Munsell Color, 1929.

———. *Munsell Book of Color, Glossy Edition*. Munsell Color, 1958.

National Institute of Science and Technology (NIST). "The Founding: Overview." NIST at 100: Foundations for Progress. NIST, U.S. Department of Commerce, Feb. 14, 2023. www.nist.gov.

"Necrology: I. H. Godlove." *Journal of the Optical Society of America* 44, no. 11 (1954): 887.

Neilson, William Allan, ed. *Webster's New International Dictionary, Second Edition*. G. & C. Merriam, 1934.

Newton, Maud. "It's Not Easy Seeing Green." *The 6th Floor* (blog), *New York Times,* Sept. 4, 2012.

Nickerson, Dorothy. "History of the Munsell Color Foundation, 1942–1972." ISCC *Newsletter* 234 (1975).

———. "History of the Munsell Color System, Company, and Foundation. II. Its Scientific Application." *Color Research and Application* 1, no. 2 (1976): 69–77. doi.org/10.1111/j.1520-6378.1976.tb00017.x.

———. "History of the Munsell Color System, Company, and Foundation. III." *Color Research and Application* 1, no. 3 (1976): 121–30. doi.org/10.1111/j.1520-6378.1976.tb00028.x.

———. *A Method for Determining the Color of Agricultural Products*. Technical Bulletin 154. Technical Bulletins. U.S. Department of Agriculture, 1929.

———. *Standardization of Color Names: The ISCC-NBS Method*. U.S. Department of Agriculture, Agricultural Marketing Service, 1940.

"Notes and Personalia." *Language* 10, no. 1 (1945): 65–68.

Nyswander, Rachel Fesler, and Janet M. Hooks. *Employment of Women in the Federal Government, 1923–1939*. Bulletin of the Women's Bureau 182. U.S. Department of Labor, Women's Bureau, 1941.

Office of Weights and Measures. "OWM Background and History." NIST Physical Measurement Laboratory. NIST, U.S. Department of Commerce, Dec. 11, 2024. www.nist.gov/pml/owm/about-us.

Ogden, C. K., and I. A. Richards. *The Meaning of Meaning*. Harcourt Brace Jovanovich, 1989.

Paclt, J. "A Chronology of Color Charts and Color Terminology for Naturalists." *Taxon* 32, no. 3 (1983): 393–405. doi.org/10.2307/1221496.

Pantone. "PANTONE 17-1230 Mocha Mousse | Pantone Color of the Year 2025." Pantone. Accessed Feb. 2, 2025. www.pantone.com.

Peek, Jenny. "I Can't Believe It's Not Yellow: A Peek into Wisconsin's Quirky Margarine Laws." *Wisconsin State Farmer*, Nov. 15, 2021. www.wisfarmer.com.

Pliny. *Natural History*. Translated by H. Rackham, W. H. S. Jones, and D. E. Eichholz. Loeb Classical Library. Vol. 10. Harvard University Press, 1952. www.attalus.org.

Porter, Noah, ed. *Webster's International Dictionary of the English Language*. G. & C. Merriam, 1898.

Priest, Irwin Gillespie, Kasson Stanford Gibson, and Harry John McNicholas. *An Examination of the Munsell Color System*. Technologic Papers of the Bureau of Standards 167. U.S. Department of Commerce, 1920.

"Program of the 1957 Winter Meeting of the Optical Society of America." *Journal of the Optical Society of America* 47, no. 4 (1957): 336–46. doi.org/10.1364/josa.47.000336.

Reingold, Nathan, and Ida H. Reingold. *Science in America: A Documentary History, 1900–1930*. University of Chicago Press, 1981.

"Reminiscences of General Aniline and Film Corporation by Harlan B. Freyermuth." Archived from original on Aug. 19, 2021. web.archive.org/web/20210819211855/http://www.colorantshistory.org/GAFFreyermuth.html.

Ridgway, Robert. *Color Standards and Color Nomenclature*. Published by the author, 1912.

Roberson, Debi, Jules Davidoff, Ian R. L. Davies, and Laura R. Shapiro. "Color Categories: Evidence for the Cultural Relativity Hypothesis." *Cognitive Psychology* 50, no. 4 (2005): 378–411. doi.org/10.1016/j.cogpsych.2004.10.001.

Roberson, Debi, Ian Davies, and Jules Davidoff. "Color Categories Are Not Universal: Replications and New Evidence from a Stone-Age Culture." *Journal of Experimental Psychology: General* 129, no. 3 (2000): 369–98. doi.org/10.1037/0096-3445.129.3.369.

Rosa, Edward B. "Expenditures and Revenues of the Federal Government." *Annals of the American Academy of Political and Social Science* 95 (1921): 1–113.

Rossiter, Margaret W. *Women Scientists in America: Before Affirmative Action, 1940–1972*. Johns Hopkins University Press, 1995.

———. *Women Scientists in America: Struggles and Strategies to 1940*. Johns Hopkins University Press, 1982.

Russell, Lindsay Rose. *Women and Dictionary-Making*. Cambridge University Press, 2018.

Ryan, Richard M., and Edward L. Deci. "Intrinsic and Extrinsic Motivations: Classic Definitions and New Directions." *Contemporary Educational Psychology* 25, no. 1 (2000): 54–67. doi.org/10.1006/ceps.1999.1020.

Science History Institute. "Fritz Haber." Accessed May 30, 2024. www.sciencehistory.org.

———. "Justus von Liebig and Friedrich Wöhler." Accessed May 30, 2024. www.sciencehistory.org.

Sears Christmas Book. Sears, Roebuck, 1952.

Servos, John W. *Physical Chemistry from Ostwald to Pauling: The Making of a Science in America*. Princeton University Press, 2021. doi.org/10.1515/9781400844180.

Skinner, David. *The Story of Ain't: America, Its Language, and the Most Controversial Dictionary Ever Published*. Harper, 2012.

Skorinko, Jeanine L., Suzanne Kemmer, Michelle R. Hebl, and David M. Lane. "A Rose by Any Other Name . . . : Color-Naming Influences on Decision Making." *Psychology and Marketing* 23, no. 12 (2006): 975–93. doi.org/10.1002/mar.20142.

Smith, A. K. "Health Hazards of the Dye Industry." *American Journal of Public Health* 10, no. 3 (1920): 255–57. doi.org/10.2105/ajph.10.3.255.

Syme, Patrick. *Werner's Nomenclature of Colours*. Reprint, Smithsonian Books, 2018.

U.S. Committee on Agriculture and Forestry. *Testimony in Regard to the Manufacture and Sale of Imitation Dairy Products*. U.S. Government Printing Office, 1886.

U.S. Pharmacopoeial Convention. *The Pharmacopeia of the United States of America (The United States Pharmacopeia)*. Easton, Pa., 1820.

Waller, Richard. "A Catalogue of Simple and Mixt Colours, with a Specimen of Each Colour Prefixt to Its Proper Name." *Philosophical Transactions of the Royal Society of London*, no. 179 (1686): 24–32.

Webster, Noah. *A Compendious Dictionary of the English Language*. Facsimile, 1st ed. Bounty Books, 1970.

Whitney, William Dwight, ed. *The Century Dictionary*. Century, 1889.

Wierzbicka, Anna. "Why There Are No 'Colour Universals' in Language and Thought." *Journal of the Royal Anthropological Institute* 14, no. 2 (2008): 407–25. doi.org/10.1111/j.1467-9655.2008.00509.x.

Wood, Francis A. "The Origin of Color-Names." *Modern Language Notes* 20, no. 8 (1905): 225. doi.org/10.2307/2917519.

Zivkovic, Bora. "The Urine Wheel." *Scientific American*. Accessed May 31, 2024. www.scientificamerican.com.

Index

Page numbers in *italics* refer to illustrations.

Abbreviations
IHG = Isaac Hahn Godlove
MW = Merriam-Webster

ILLUSTRATION CREDITS

The Ogden-Richards Raccoon (page 14) was illustrated and designed by Ansa Stamper.

The Munsell solid (page 34) was drawn by a nameless artist at the Munsell Color Company for Dorothy Nickerson, who commissioned it for the guide to her 1957 *Nickerson Color Fan*. It was originally a black-and-white line drawing; it was sent to the Beck Engraving Company, which did the color plates for Merriam-Webster for the *Third,* and colored possibly by the Munsell Color Company, though there's no record of such a thing happening, or by Beck's in-house artists. In either case, Dorothy Nickerson appears to have supplied the specifications for the color work: There is a note in a letter to Deane Judd that she thought the colors on the Merriam plate turned out pretty okay, considering the back-and-forth she had with their printers about the color tolerances.

The two Munsell arrays (page 257) showing the Himba and English subdivisions for *dumbu* and *burou* and blue and green, respectively, appeared in Roberson, Davies, and Davidoff (2000) in black and white and were drawn by the authors. The author superimposed their charts on her own color copy of the Munsell array.

Similarly, the author composed the two illustrations comparing *burou* to *burou* (page 258) and *dumbu* to *burou* (page 259). All complaints about these should be sent directly to the author, preferably in all caps.

A NOTE ABOUT THE AUTHOR

KORY STAMPER is a lexicographer who spent almost two decades writing dictionaries at Merriam-Webster and is currently a defining consultant for the Colour Literacy Project. She is the author of *Word by Word*. Her writing has appeared in *The Guardian, The New York Times, New York,* and *The Washington Post,* and she blogs regularly on language and lexicography at www.korystamper.com.

A NOTE ON THE TYPE

This book was set in Monotype Dante, a typeface designed by Giovanni Mardersteig (1892–1977). Conceived as a private type for the Officina Bodoni in Verona, Italy, Dante was originally cut for hand composition by Charles Malin, the famous Parisian punch cutter, between 1946 and 1952. Its first use was in an edition of Boccaccio's *Trattatello in laude di Dante* that appeared in 1954. The Monotype Corporation's version of Dante followed in 1957. Although modeled on the Aldine type used for Pietro Cardinal Bembo's treatise *De Aetna* in 1495, Dante is a thoroughly modern interpretation of the venerable face.

Composed by North Market Street Graphics,
Lancaster, Pennsylvania

Designed by Cassandra J. Pappas